CISM COURSES AND LECTURES

The series presents lecture notes, monographs, edited works and proceedings in the field of Mechanics, Engineering, Computer Science and Applied Mathematics.
Purpose of the series in to make known in the international scientific and technical community results obtained in some of the activities organized by CISM, the International Centre for Mechanical Sciences.

INTERNATIONAL CENTRE FOR MECHANICAL SCIENCES

COURSES AND LECTURES - No. 345

PASSIVE AND ACTIVE STRUCTURAL VIBRATION CONTROL IN CIVIL ENGINEERING

EDITED BY

T.T. SOONG AND M.C. COSTANTINOU
STATE UNIVERSITY OF NEW YORK AT BUFFALO

Springer-Verlag Wien GmbH

Le spese di stampa di questo volume sono in parte coperte da
contributi del Consiglio Nazionale delle Ricerche.

This volume contains 202 illustrations

In order to make this volume available as economically and as
rapidly as possible the authors' typescripts have been
reproduced in their original forms. This method unfortunately
has its typographical limitations but it is hoped that they in no
way distract the reader.

ISBN 978-3-211-82615-7   ISBN 978-3-7091-3012-4 (eBook)
DOI 10.1007/978-3-7091-3012-4

PREFACE

In structural engineering, one of the constant challenges is to find new and better means of designing new structures or strengthening existing ones so that they, together with their occupants and contents, can be better protected from the damaging effects of destructive environmental forces such as wind, wave loads, and earthquakes. As a result, new and innovative concepts of structural protection have been advanced and are at various stages of development. Structural protective systems can be divided into three groups as shown in Table I. The technique of seismic isolation is now widely used in many parts of the world. A seismic isolation system is typically placed at the foundation of a structure which, by means of its flexibility and energy absorption capability, partially absorbs and partially reflects some of the earthquake input energy before it is transmitted to the structure. The net effect is a reduction of energy dissipation demand on the structural system, resulting in an increase in its survivability.

Table I. Structural Protective Systems

Seismic Isolation	Passive Energy Dissipation	Active Control
Elastomeric Bearings	Metallic Dampers	Active Bracing Systems
Lead Rubber Bearings	Friction Dampers	Active Mass Dampers
Elastomeric Bearings with Energy Dissipating Devices	Viscoelastic Dampers	Active Variable Stiffness or Damping Systems
Sliding Friction Pendulum	Viscous Dampers	Pulse Systems
Flat Sliding Bearings with Restoring Force Devices	Tuned Mass Dampers	Aerodynamic Appendages
Lubricated Sliding Bearings with Energy Dissipating Devices	Tuned Liquid Dampers	

Much progress has also been made in research and development of passive energy dissipation devices for structural applications. Similar to seismic isolation technology, the basic role of passive energy dissipation devices when incorporated into a structure is to absorb or consume a portion of the input energy, thereby reducing energy dissipation demand on primary structural members and minimizing possible structural damage. Unlike seismic isolation, however, these devices can be effective against wind excited motions as well as those due to earthquakes. In recent years, serious efforts have been undertaken to develop the concept of energy dissipation or supplemental damping into a workable technology, and a number of these devices have been installed in structures throughout the world.

Active control research has a more recent origin. Active structural control is an area of structural protection in which the motion of a structure is controlled or modified by means of the action of a control system through some external energy supply. Considerable attention has been paid to active structural control research in recent years. It is now at the stage where actual systems have been designed, fabricated and installed in full-scale structures.

This collection of lecture notes represents an attempt to introduce the basic concepts of these relatively new technologies, to provide a working knowledge of this exciting and fast expanding field, and to bring up-to-date current research and world-wide development in seismic isolation, passive energy dissipation, and active control. The book is divided into three parts, each addressing one of these topics. In each case, basic principles are introduced, followed by design and applications, implementation issues, case studies, and code issues if applicable.

It is a great pleasure to acknowledge the significant contributions made to this lecture series by Professor Peter Hagedorn of the Technische Hoshschule Darmstadt, Germany; Professor Hirokazu Iemura of Kyoto University, Japan; and Professor Jose Rodellar of the Universidad Politecnica de Catalunya, Spain. They delivered excellent lectures at Udine in June, 1993, and contributed important chapters in this book, making this project a truly international effort.

T.T. Soong
M.C. Constantinou

CONTENTS

Page

CHAPTER I

MECHANICAL VIBRATIONS AND VIBRATION CONTROL

P. Hagedorn

Technical University of Darmstadt, Darmstadt, Germany

Abstract

In these lectures, an introduction is given into the basic concepts of the theory of active and passive vibration damping. First one degree of freedom systems are discussed with respect to the different solution techniques and with view to damping and isolation mechanisms. Next the theory of optimal control is briefly introduced and some examples are given. Mechanical systems with n degrees of freedom are discussed and the concepts of mechanical impedance, vibration absorption and impedance matching are studied. The sources of nonlinearities in structural vibrations are examined and an appropriate solution technique is explored. Finally, special attention is devoted to a particular technical application, namely to the wind-exited vibrations of transmission lines, where the problem of the damping of bundles of conductors is discussed in more detail.

1 General Remarks on Damping, Systems with One Degree of Freedom

1.1 Introduction: Systems with One Degree of Freedom

In what follows we shall briefly review the forced oscillations of a one-degree of freedom system described by

$$m\ddot{x} + d\dot{x} + cx = f(t) \qquad (1.1)$$

where $f(t)$ is a given forcing function. The coefficients m, d, and c respectively stand for mass, damping, and stiffness.

It is often convenient to introduce the parameters $\omega := \sqrt{c/m}, \tau = \omega t$. Indicating the derivatives with respect to the new dimensionless time τ by a prime equation (1.1) can be reduced to

$$x''(\tau) + 2Dx'(\tau) + x(\tau) = \frac{1}{m}f\left(\frac{\tau}{\omega}\right),\tag{1.2}$$

where

$$D = \frac{d}{2\sqrt{cm}}\tag{1.3}$$

is the non-dimensional damping parameter which in most structures assumes values between 0.001 and 0.02.

1.2 Forced Oscillations

1.2.1 Response to Harmonic Excitation, Isolation

The case of harmonic force excitation

$$f(t) = \hat{f}\cos(\Omega t + \alpha)\tag{1.4}$$

or in complex notation

$$f(t) = Re\underline{f}(t)\tag{1.5}$$

with

$$\underline{f}(t) = \hat{f}e^{j\alpha}e^{j\Omega t} = \underline{\hat{f}}e^{j\Omega t}\tag{1.6}$$

is the most elementary one. In our notation we use $\hat{\ }$ to indicate "amplitude" and underline complex quantities. The character j is used to denote the imaginary unit. The general solution to (1.1) is of the type

$$x(t) = x_H(t) + x_P(t).\tag{1.7}$$

In many applications we are interested mainly in $x_P(t)$, that is in the particular solution of the inhomogenous equation. In the case of forced harmonic oscillations the *Ansatz*

$$\underline{x}_P(t) = \underline{\hat{x}}_P e^{j\Omega t}\tag{1.8}$$

leads to

$$\underline{\hat{x}}_P = \underline{G}(\Omega)\underline{\hat{f}},\tag{1.9}$$

where the transfer function $\underline{G}$ is the complex dynamic compliance of the system, and it is given by the function

$$\underline{G}(\Omega) := \frac{1}{c} \frac{1}{1 - \left(\frac{\Omega}{\omega}\right)^2 + j\frac{2D\Omega}{\omega}} = \frac{1}{c} \frac{1}{1 - \eta^2 + j2D\eta} \tag{1.10}$$

with $\eta := \Omega/\omega$. The magnitude and the phase of the transfer function are respectively given by

$$|\underline{G}(\eta)| = \frac{1}{c} \frac{1}{\sqrt{(1 - \eta^2)^2 + (2D\eta)^2}}, \tag{1.11}$$

$$\tan[\arg \underline{G}(\eta)] = -\frac{2D\eta}{1 - \eta^2}. \tag{1.12}$$

As we can see in fact $\underline{G}$ has the dimensions of an inverse stiffness, so that the function

$$V_A(\eta) = c|\underline{G}(\eta)| \tag{1.13}$$

is non-dimensional. This function is shown in Fig. 1.1.

The function $V_A(\eta)$ assumes a particularly simple form in double logarithmic representation since it can be approximated by straight lines for all points which are far away from the resonance $\eta = 1$. The maximum of V_A depends on D and is nearly equal to $1/(2D)$.

Of course the case of a given harmonic force exciting a structure is not really very common, typically the exciting forces act through elastic or visco-elastic elements on the oscillating system. In many cases kinematical variables, such as displacement or velocities are given as functions of the time instead of the forces. This is of course the case for a structure excited through base motion. We briefly consider the different types of harmonic excitation shown in Fig. 1.2.

In the first case shown in Fig. 1.2a the end-point of the spring moves according to

$$x_F(t) = \hat{x}_F \cos \Omega t. \tag{1.14}$$

This gives an equation of motion of the type

$$m\ddot{x} + d\dot{x} + cx = cx_F(t) = c\hat{x}_F \cos \Omega t. \tag{1.15}$$

In the second case the damper end-point moves according to

$$x_D(t) = \hat{x}_D \cos(\Omega t - \pi/2) \tag{1.16}$$

giving an equation of motion

$$m\ddot{x} + d\dot{x} + cx = d\dot{x}_D(t) = d\Omega\hat{x}_D \cos \Omega t. \tag{1.17}$$

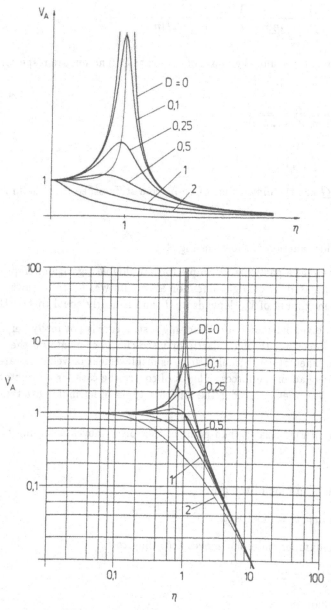

Figure 1.1: The function $V_A(\eta)$ in linear and logarithmic representation

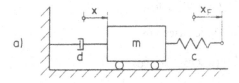

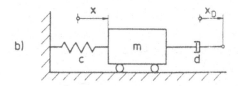

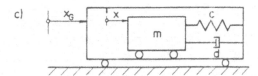

Figure 1.2: Different types of harmonic excitation:
a) given displacement of spring end-point
b) given displacement of spring damper
c) base excitation

Finally in Fig. 1.2c the base motion is described by

$$x_G(t) = \hat{x}_G \cos \Omega t \tag{1.18}$$

which leads to

$$m\ddot{x} + d\dot{x} + cx = -m\ddot{x}_G(t) = m\Omega^2 \hat{x}_G \cos \Omega t. \tag{1.19}$$

Note that in the last case the coordinate x is the relative displacement of the mass m with respect to the moving base. In all three cases in steady-state the amplitude $\hat{x}$ is equal to the excitation amplitude of the given displacement multiplied by an amplification factor. In the case of Fig. 1.2a the amplification factor is exactly the function $V_A(\eta)$ shown in Fig. 1.1.

In the other two cases the amplification factors can be written as

$$V_B(\eta) = 2D\eta V_A(\eta), \ V_C(\eta) = \eta^2 V_A(\eta),$$ (1.20)

so that the amplitudes in the three cases are

$$\hat{x}_{P1} = V_A\hat{x}_F, \ \hat{x}_{P2} = V_B\hat{x}_D, \ \hat{x}_{P3} = V_C\hat{x}_G.$$ (1.21)

Figs. 1.3a and 1.3b show the amplification factors V_B and V_C.

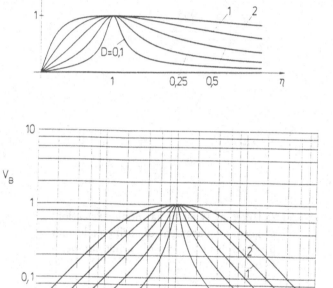

Figure 1.3a: The function $V_B(\eta)$ in linear and logarithmic representation

So far we have only looked at the amplitudes of the displacements. Frequently, of course, we are interested in the amplitudes of the velocities or of the accelerations which can easily be obtained by differentiation of this stationary solution. Each differentiation implies a multiplication of the amplification factor by η.

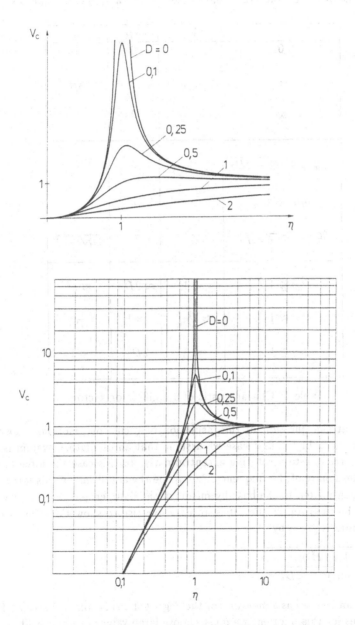

Figure 1.3b: The function $V_C(\eta)$ in linear and logarithmic representation

In all three cases for small values of the damping ratio the maximum values of the amplification factor with respect to the dimensionless frequency is approximately equal to $1/(2D)$. Table 1 gives some of the characteristic parameters of the amplification factors.

η	V_A	V_B	V_C
0	1	0	0
1	$\frac{1}{2D}$	1	$\frac{1}{2D}$
∞	0	0	1
$\eta_{Amax} = \sqrt{1 - 2D^2}$	$\frac{1}{2D\sqrt{1-D^2}}$	-	-
$\eta_{Bmax} = 1$	-	1	
$\eta_{Cmax} = \frac{1}{\sqrt{1-2D^2}}$	-	-	$\frac{1}{2D\sqrt{1-D^2}}$
$\eta \leq 1$	≈ 1	$\approx 2D\eta$	$\approx \eta^2$
$\eta \approx 1$	$\approx \frac{1}{2D}\frac{1}{\eta}$	≈ 1	$\frac{1}{2D}\eta$
$\eta \geq 1$	$\approx \frac{1}{\eta^2}$	$2D\frac{1}{\eta}$	≈ 1

Table 1: Characteristic of amplification factors

Let us now look at the question of the insulation of oscillatory motions. Two different aspects can be studied. In the first case we assume that an oscillatory system is subjected to the harmonic force due to a rotating excentric mass. In this case the force amplitude is proportional to the square of the frequency. Let $\hat{f}_U$ be the amplitude of this force. In steady state the forces transmitted to the base by means of the damper and the spring will also be harmonic, let $\hat{f}_Z$ be their amplitude. It is an easy exercise to show that the ratio of these two force amplitudes is given by

$$\frac{\hat{f}_Z}{\hat{f}_U} = \sqrt{\frac{1 + (2D\eta)^2}{(1 - \eta^2)^2 + (2D\eta)^2}}. \tag{1.22}$$

The ratio $\hat{f}_Z/\hat{f}_U$ can be used as a measure for the degree of insulation in this case. In order to obtain small values for this quotient we must choose large values of η (i.e. soft springs) and small values of D. However, if D is too small, there may be dangerously large amplitudes during the process of speeding up the rotor in the neighborhood of the resonance $\eta \approx 1$.

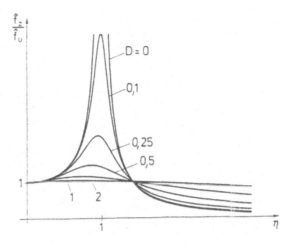

Figure 1.4: The ratio $\hat{f}_Z / \hat{f}_U$

Therefore a compromise has to be achieved. Possibly some damping can be provided during the starting of the rotor which can later be "turned off".

Let us now look at the problem of isolating the motion of a rigid body against base excitation. Frequently in this case the absolute acceleration of the base-excited mass is to be kept as small as possible. This is not the variable x in Fig. 1.2c but rather $x + x_G$. The quotient of the complex amplitudes of the body's displacement and the base displacement can be used as a measure of the quality of vibration insulation. It turns out that exactly the same expression (1.22) characterizes this quotient, so that the same results are obtained. Protecting a structure or elastically supported machine from base excitation therefore is equivalent to minimizing the forces applied by the structure to the base in the case of excitation due to an unbalanced rotating mass.

1.2.2 Impact Response

The response to a harmonic force excitation described in the previous section completely characterizes the system, i.e. if the transfer function $\underline{G}$ is known, the response to any input $f(t)$ can be computed. This, however, only one way of characterizing a linear system. The system's properties may also be described via the response to other functions, such as the unit step function or the unit impulse.

Consider equation (1.1) with $f(t) = s(t)$, the unit step function. It is a simple exercice to compute the response

$$h(t) = \frac{1}{c}\left[1 - e^{-\delta t}\left(\cos \omega_d t + \frac{\delta}{\omega_d}\sin \omega_d t\right)\right] s(t) \qquad (1.23)$$

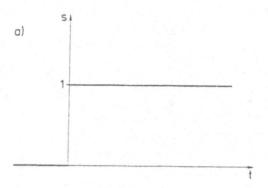

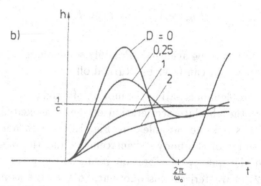

Figure 1.5: Unit step function $s(t)$ and step response $h(t)$

to the unit step function $s(t)$ with

$$\delta = \omega D, \quad \omega_d = \omega\sqrt{1 - D^2}. \tag{1.24}$$

These expressions are valid only for $D < 1$. The unit step function $s(t)$ and the step response $h(t)$ are shown in Fig. 1.5.

A third way of characterizing the system is through the response of the unit impulse, i.e. the Dirac function $f(t) = \delta(t)$. The impulse response can be written as

$$g(t) = \frac{s(t)}{m\omega_d} e^{-\delta t} \sin \omega_d t \tag{1.25}$$

for $D < 1$ and is shown in Fig. 1.6.

It is easy to verify that $\delta(t)$ and $g(t)$ are respectively the generalized time derivatives of $s(t)$ and $h(t)$.

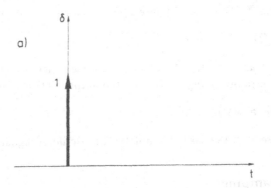

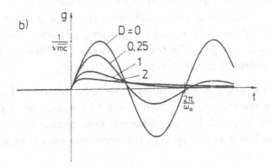

Figure 1.6: Unit impact and impact response

If the FOURIER transform is applied to $g(t)$ we obtain

$$g(t) \rightarrow \underline{G}(\Omega), \tag{1.26}$$

i.e. the previously considered transfer function is exactly the FOURIER transform of the impact response.

1.2.3 Response to Arbitrary Forces

Let us now consider briefly equation (1.1) for arbitrary time functions $f(t)$. If we consider the FOURIER transform $\underline{F}(\Omega)$ of $f(t)$,

$$f(t) \rightarrow \underline{F}(\Omega), \tag{1.27}$$

the FOURIER transform $\underline{X}(\Omega)$ of the response $x(t)$ can be computed by

$$\underline{X}(\Omega) = \underline{G}(\Omega)\underline{F}(\Omega). \tag{1.28}$$

The inverse FOURIER transformation then gives the time function $x(t)$. Also the relation between the power spectra of the input $f(t)$ and the output $x(t)$ is often used:

$$R_{xx}(\Omega) = |\underline{G}(\Omega)|^2 \, R_{ff}(\Omega). \tag{1.29}$$

This formula is also useful in the case of non-deterministic input signals.

1.3   Structural Damping

So far in equation (1.1) we have considered linear springs, ideal linear dampers, and a rigid mass. In reality damping and restoring forces can often not be physically separated, both damper and spring forming one physical element. This is the case, e.g. for rubber springs, where the material law may contain high damping terms and also in disc springs, where mechanical energy is dissipated by friction between the different discs. Fig. 1.7 shows symbolically different models of massless supporting elements.

The damping properties of real systems usually have to be determined experimentally. This can be done submitting the material or the element to a harmonic displacement and measuring the corresponding force. A hysteresis loop is then obtained in the f-x-plane. If the element is linearly viscoelastic as in Fig. 1.7a the hysteresis loop is an ellipse with the area

$$\Delta E = \pi d\Omega \hat{x}_P^2 \tag{1.30}$$

equal to the energy loss for one cycle.

In (1.30) the energy loss per cycle is proportional to the square of the displacement amplitude and to the frequency. In the corresponding elliptic hysteresis loop of Fig. 1.8a the line defining the linear spring characteristic is very close to an axis of symmetry of the ellipse.

Real materials and structures usually, however, do not present linear viscoelastic behavior, and the hysteresis loops have different forms, such as, e.g. in Fig. 1.8b. Each hysteresis loop can usually be approximated by an ellipse, however, the parameters of this ellipse do not correspond to a linear viscoleastic law if the frequency and amplitude are changed. Experience shows that the energy loss in many structures is quadratic in the amplitude $\hat{x}_P$ but at the same time is almost independent on the frequency Ω.

In these cases the energy loss per cycle can be approximated by

$$\Delta E = k\hat{x}_P^2, \tag{1.31}$$

where k is a constant, rather than by (1.30). The dissipation of mechanical energy in these cases is usually due to a variety of simultaneous phenomena, where nonlinear material laws

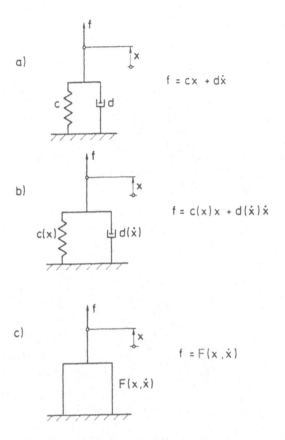

Figure 1.7: Different models of massless supporting elements
a) Linear-viscoelastic
b) Combination of nonlinear spring with nonlinear damper
c) General element

usually only play a secondary role. Most of the mechanical energy is typically dissipated at the boundaries and joints of parts of the structure. In mechanical engineering structures it is well known that e.g. welded structures present much lower damping than bolted or riveted structures. In civil engineering structures the situation is similar.

This type of damping, described by (1.31) is called *structural damping*. It can be also described by defining a damping constant in equation (1.1) which is inversely proportional to the frequency:

$$d = \frac{k}{\pi \Omega}. \tag{1.32}$$

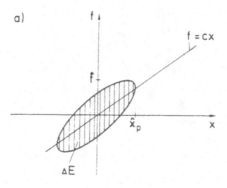

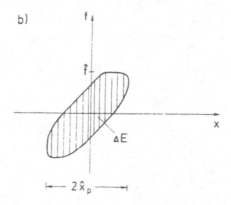

Figure 1.8: Hysteresis loops

The equation of motion in complex notation has then the form

$$m\underline{\ddot{x}} + \frac{k}{\pi\Omega}\underline{\dot{x}} + c\underline{x} = \hat{\underline{f}}e^{j\Omega t}$$

(1.33)

in the case of a harmonic force excitation. This gives rise to the transfer function

$$\underline{G_D}(\Omega) = \frac{1}{c} \frac{1}{1 - \left(\frac{\Omega}{\omega}\right)^2 + j\frac{k}{\pi c}}$$

(1.34)

The quantity $\tan \nu = k/(\pi c)$ is the *loss factor*, which is equal to the tangent of the loss angle. The loss angle is often used to characterize material damping, e.g. of rubber-like materials.

Exactly the same result is obtained if we substitute (1.33) by

$$m\underline{\ddot{x}} + c(1 + j\tan\nu)\underline{x} = \hat{\underline{f}}e^{j\Omega t}.$$

(1.35)

In this case the damping is contained in the imaginary part of the "complex stiffness" $c(1 + j \tan \nu)$. This formulation is particularly suitable for continuous systems, where the damping can be introduced a posteriori into the conservative system by substituting the modulus of elasticity E by the "complex modulus of elasticity" $E(1 + j \tan \nu)$.

It is not clear how this structural damping law should be applied to a system under an arbitrary force excitation $f(t)$. One idea would of course be to use the transfer function (1.34) for the determination of the response to an arbitrary force input according to (1.28). We noted previously that the inverse FOURIER transform of the transfer function is equal to the impact response. The inverse transform of (1.34) is, however, not a real function! This follows from the fact that the FOURIER transforms of real functions always are hermitean (they have an even real and an odd imaginary part), which is not the case for (1.34). Moreover, a more careful examination shows that the inverse transform of (1.34) is not even a causal function, i.e. the impact response is present before the impact!

These considerations show that structural damping according to (1.32) is highly unsatisfactory from a mathematical point of view, although real structures are usually well characterized in this manner in the case of harmonic forcing functions. Of course the frequency range of interest in structural vibrations is usually quite limited, so that also the transfer function is defined by (1.34) in a very limited frequency band only. For nonharmonic excitation it may be essential to take into account nonlinearities in the damping law if high precision is required.

References:

Hagedorn P., Otterbein S.: Technische Schwingungslehre, Springer, Berlin 1987.

2   Some Remarks on the Theory of Optimal Control

2.1   Control Problems, Controllability

Since this course deals with passive and active damping of structures, a short introduction to control theory would be in place. In this lecture we will limit ourselves mainly to a few remarks concerning the theory of *optimal control*. Consider equation (1.1) of lecture 1 and suppose that the force $f(t)$ on the right-hand side is to be chosen as a function of time t in such a way as to damp the vibrations, steering the sytem to equilibrium in minimal time. The determination of this function $f(t)$ is then typically an exercise in optimal control.

In control theory it usually is convenient to write the equations of motion in the form of a first-order system. Moreover, in what follows we will not restrict ourselves to the one degree of freedom oscillator but we will treat somewhat more general problems. Examples will be given not only related to the active vibration damping but also to other fields.

Consider a system of the form

$$\dot{\mathbf{x}} = \mathbf{f}(\mathbf{x}, \mathbf{u}, t) \tag{2.1}$$

with $\mathbf{x} \in R^n$, $\mathbf{u} \in R^m$. With the introduction of a given function of time $\mathbf{u}(t)$ into the right-hand side of equation (2.1), one obtains an ordinary, generally non-linear, differential equation. In many physical processes described by differential equations, the time dependence of the process may be influenced in some manner. This is generally referred to as a steering of the process: it is implemented in equation (2.1) by means of the *steering function* or *control function* $\mathbf{u}(t)$. The variable $\mathbf{x}$ is called the *state variable* in control theory. Consider, for example, a spaceship; then the components of the vector $\mathbf{x}$ may refer to position and velocity and $\mathbf{u}^T = (u_1, u_2, \ldots, u_m)$ may represent the controllable thrusts of the individual engines as well as their directions of action insofar as the engines are gimballed. If a system is to be transferred from a given state $\mathbf{x}_0$ to another state $\mathbf{x}_1$ there often exist numerous control functions with which this may be accomplished. Not all of these will be suitable and the question arises how one might determine a most suitable one - for example, that one which minimizes the fuel consumption, or that which minimizes the travel time.

The solutions of the system (2.1) depend on the function $\mathbf{u}(\cdot)$ and one often writes them in the form $\mathbf{x}(\mathbf{x}_0, t_0; \mathbf{u}(\cdot), t)$. A typical question in control theory then is the following: For a given $\mathbf{x}_0$, t_0, and $\mathbf{x}_1$, does there exist a $t_1 > t_0$ and a function $\mathbf{u}(\cdot)$ such that $\mathbf{x}(\mathbf{x}_0, t_0; \mathbf{u}(\cdot), t_1) = \mathbf{x}_1$, that is, does there exist a function $\mathbf{u}(\cdot)$ which steers the system from $\mathbf{x} = \mathbf{x}_0$ to $\mathbf{x} = \mathbf{x}_1$? If the answer to this question is affirmative, then one also says that the system (2.1) is *controllable* from $\mathbf{x}_0$ to $\mathbf{x}_1$.

An important special case is that of the linear system with constant coefficients:

$$\dot{\mathbf{x}} = \mathbf{A}\mathbf{x} + \mathbf{B}\mathbf{u}, \tag{2.2}$$

where $\mathbf{A}$ is an $n \times n$ matrix and $\mathbf{B}$ is an $n \times m$ matrix. For linear systems, it is meaningful to define *complete controllability*:

The system (2.2) is completely controllable if and only if for every $\mathbf{x}_0$ there exists a function $\mathbf{u}(\cdot)$ which steers the system to $\mathbf{x}_1 = \mathbf{0}$.

For the system (2.2) there exists a simple criterion for controllability. The system (2.2) is completely controllable if and only if

$$\text{rank} \, (\mathbf{B}, \mathbf{A}\mathbf{B}, \ldots, \mathbf{A}^{n-1}\mathbf{B}) = n \tag{2.3}$$

holds.

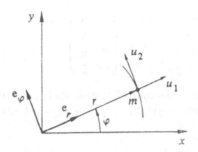

Figure 2.1: Satellite in a gravitational field

Examples:

(1) Satellite in a central gravitational field. With kinetic energy $T = \frac{1}{2}m(\dot{r}^2 + r^2\dot{\varphi}^2)$ and potential energy $U = -\gamma m/r$, the equations of motion are obtained as

$$\left.\begin{aligned} \ddot{r} &= -\frac{\gamma}{r^2} + r\dot{\varphi}^2, \\[2mm] \ddot{\varphi} &= -2\frac{\dot{\varphi}\dot{r}}{r}. \end{aligned}\right\} \tag{2.4}$$

Here m is the mass of the satellite, γ is the product of the gravitational constant with the mass of the planet, and r and φ are the polar coordinates of the satellite, idealized as a particle, in the xy-plane (Fig. 2.1). A particular solution is given by the motion in a circular orbit with $r = R, \dot{\varphi} = \omega = \sqrt{(\gamma/R^3)}$. Deviations from this orbit are to be corrected by two rocket engines with thrust vectors in the directions $\vec{e_r}$ und $\vec{e_\varphi}$, respectively, so that the equations of motion now have the form

$$\left.\begin{aligned} \ddot{r} &= r\dot{\varphi}^2 - \frac{\gamma}{r^2} + u_1, \\[2mm] \ddot{\varphi} &= -2\frac{\dot{\varphi}\dot{r}}{r} + \frac{1}{r}u_2, \end{aligned}\right\}$$

where u_1 und u_2 are the accelerations in the radial and normal directions, respectively, as produced by the corresponding engines. The expressions "radial" and "tangential" here refer to the circular orbit. The introduction of the new variables

$$\left.\begin{aligned} x_1 &= r - R, \\ x_2 &= \dot{x}_1 = \dot{r}, \\ x_3 &= (\varphi - \omega t)R, \\ x_4 &= \dot{x}_3 = (\dot{\varphi} - \omega)R, \end{aligned}\right\} \tag{2.5}$$

which represent the (small) deviations from the reference motion, and the subsequent linearization lead to differential equations of the form (2.2) with

$$
\mathbf{A} = \begin{pmatrix} 0 & 1 & 0 & 0 \\ 3\omega^2 & 0 & 0 & 2\omega \\ 0 & 0 & 0 & 1 \\ 0 & -2\omega & 0 & 0 \end{pmatrix}, \qquad \mathbf{B} = \begin{pmatrix} 0 & 0 \\ 1 & 0 \\ 0 & 0 \\ 0 & 1 \end{pmatrix}, \tag{2.6}
$$

providing a first approximation to the controlled motions in a neighbourhood of the reference orbit. To investigate the controllability of the system, the rank of $(\mathbf{B}, \mathbf{AB}, \mathbf{A}^2\mathbf{B}, \mathbf{A}^3\mathbf{B})$ must be computed. One has

$$
\mathrm{rank}(\mathbf{B}, \mathbf{AB}, \mathbf{A}^2\mathbf{B}, \mathbf{A}^3\mathbf{B}) =
$$

$$
= \mathrm{rank} \begin{pmatrix} 0 & 0 & 1 & 0 & 0 & 2\omega & -\omega^2 & 0 \\ 1 & 0 & 0 & 2\omega & -\omega^2 & 0 & 0 & -2\omega^3 \\ 0 & 0 & 0 & 1 & -2\omega & 0 & 0 & -4\omega^2 \\ 0 & 1 & -2\omega & 0 & 0 & -4\omega^2 & 2\omega^3 & 0 \end{pmatrix} = 4,
$$

and it follows that the system indeed is completely controllable.

If one assumes that only radial thrust is available, then $\mathbf{B}^T = (0, 1, 0, 0)$ follows and the system no longer is completely controllable, as may easily be checked. If one, however, admits only tangential thrust with $\mathbf{B}^T = (0, 0, 0, 1)$, the complete controllability is maintained.

(2) *The concept of pervasive damping.* The linear oscillations of a damped discrete mechanical system about a stable equilibrium position are described by

$$
\mathbf{M}\ddot{\mathbf{x}} + \mathbf{D}\dot{\mathbf{x}} + \mathbf{C}\mathbf{x} = 0 \tag{2.7}
$$

where $\mathbf{M}^T = \mathbf{M}$ and $\mathbf{C}^T = \mathbf{C}$ are positive definite constant matrices. If the damping matrix $\mathbf{D}^T = \mathbf{D}$ also is positive definite, then (2.7) is "completely damped" and the trivial solution is asymptotically stable. This is a sufficient but by no means necessary condition for asymptotic stability. More specifically, the system may be asymptotically stable even when $\mathbf{D}$ is only semi-definite; the latter case is referred to as *pervasive damping*. It is easy to show that the complete controllability of the system

$$
\mathbf{M}\ddot{\mathbf{x}} + \mathbf{D}\mathbf{u} + \mathbf{C}\mathbf{x} = 0 \tag{2.8}
$$

(written as a first-order system) together with the positive semi-definiteness of $\mathbf{D}$ is a necessary and sufficient condition for asymptotic stability. The transformation of the system (2.8) to the principal coordinates of the undamped system results in

$$
\ddot{y}_i + \omega_i^2 y_i + \mathbf{l}_i^T \mathbf{u} = 0, \quad i = 1, 2, \ldots, n, \tag{2.9}
$$

where the $\mathbf{l}_i \in R^m$ are constant vectors. If the vector $\mathbf{l}_i$ vanishes for some i then the corresponding subsystem (2.9) is not completely controllable and the corresponding eigenmode is undamped.

2.2 The Pontryagin Maximum Principle

Consider again the system (2.1) and assume that it is desired to transfer the system from the given state $\mathbf{x}_0$ to the terminal state $\mathbf{x}_1$, which is also specified. Furthermore, the control function $\mathbf{u}(t)$ is assumed to be piecewise continuous and it is to take on values in the so-called "control set" U. A solution of the control problem then is a pair of functions $\mathbf{u}(t), \mathbf{x}(t)$ satisfying equations (2.1), with $\mathbf{u}(t) \in U, t_0 \leq t_1$ and $\mathbf{x}(t_0) = \mathbf{x}_0, \mathbf{x}(t_1) = \mathbf{x}_1$. The corresponding function $\mathbf{u}(t)$ is called an admissible control function. The instants t_0 and t_1 here generally have not been specified *a priori*.

The control set U is assumed to be a closed bounded set as, for example, the "unit ball" $\| \mathbf{u} \| \leq 1$ or the "unit sphere" $\| \mathbf{u} \| = 1$; but the control set may also be specified in terms of several equations or inequalities of the form $g(\mathbf{u}, \mathbf{x}, t) \leq 0$. In applications these constraints arise from simple physical limitations imposed on the acceleration, velocity, or force, for example.

The question concerning the existence of a solution for the control problem formulated above will not be treated here. Rather, it will be assumed that solutions exist and necessary conditions for the "optimal" solution will be formulated. An optimal solution of the control problem here is one for which the "cost functional"

$$J = \int_{t_0}^{t_1} f_0(\mathbf{x}(t), \mathbf{u}(t), t) dt \qquad (2.10)$$

takes on its smallest possible value. Here, f_0 is a given function of $\mathbf{x}$, $\mathbf{u}$, and t. Note again, that t_1 is not specified but that it depends on whichever control function $\mathbf{u}(t)$ is chosen. In particular, f_0 may be identically equal to unity so that the optimal solution for the control problem is that which yields the shortest transit time from $\mathbf{x}_0$ to $\mathbf{x}_1$ (called "time-optimal control").

The optimal control problem may also be formulated in a somewhat different manner by introducing the additional variable x_0[1] and dealing with the system

$$\dot{x}_i = f_i(x_1, \ldots, x_n, \mathbf{u}, t), \quad i = 0, 1, 2, \ldots, n. \qquad (2.11)$$

The given quantities are assumed to be the point $(0, x_1(t_0), x_2(t_0), \ldots, x_n(t_0))$ in R^{n+1} as well as the straight line $x_1(t_1), x_2(t_1), \ldots, x_n(t_1)$ parallel to the x_0-axis, where t_1 is yet undetermined. Determine that control function $\mathbf{u}(t)$ whose corresponding solution of equations (2.11) cuts the straight line as close as possible to the hyperplane $x_0 = 0$ (Fig. 2.2).

Necessary conditions for the optimality of a solution of the control problem are supplied by the Pontryagin *maximum principle*, which is cited here without proof.

---

[1] The additional variable x_0 should not be confused with the initial state $\mathbf{x}(t_0) = \mathbf{x}_0$

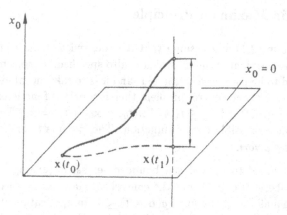

Figure 2.2: The problem of optimal control

Let $(\mathbf{u}(t), \mathbf{x}(t))$ be a solution of the boundary-value problem defined by the system (2.1), $\mathbf{x}_0$ and $\mathbf{x}_1$ with free terminal time, and minimizing the functional (2.10). Then there exist continuous piecewise differentiable functions $y_0(t), y_1(t), \ldots, y_n(t)$ which satisfy the following conditions:

a) The functions $\mathbf{x}_1(t)$ which satisfy the differential equations

$$\dot{x}_i = \frac{\partial H}{\partial y_i}, \tag{2.12}$$

$$\dot{y}_i = -\frac{\partial H}{\partial x_i}, \quad i = 0, 1, 2, \ldots, n, \tag{2.13}$$

with

$$H(x_1, \ldots, x_n; y_0, \ldots, y_n; \overline{\mathbf{u}}(t), t) =$$
$$= \sum_{i=0}^{n} y_i f_i(x_1, \ldots, x_n; \overline{\mathbf{u}}(t), t); \tag{2.14}$$

b) the $y_0, y_1, \ldots, y_n$ are not all simultaneously zero, and y_0 is a constant ≤ 0;

c) one has

$$H(x_1(t), \ldots, x_n(t); y_0(t), \ldots, y_n(t); \overline{\mathbf{u}}(t), t) =$$
$$= \max_{\mathbf{u} \in U} H(x_1(t), \ldots x_n(t); y_0(t), \ldots, y_n(t); \mathbf{u}(t), t), \tag{2.15}$$

that is, the optimal control function at each instant assumes those values which maximize H;

d) in terms of the optimal solution, H is a continuous function of the time and one has $H(t_1) = 0$. If the functions $f_i, i = 0, 1, 2, \ldots, n$ do not depend on t explicitly, then H is a constant and equal to zero.

Clearly, equations (2.12) and (2.13) describe Hamiltonian systems where the x_i correspond to the generalized coordinates and where the additionally introduced adjoint variables y_i correspond to the generalized momenta. Recall also that the Hamiltonian is a first integral for autonomous systems, taking on the value zero in the case of an optimal control, as indicated by the assertion (d) of the theorem.

If (2.1), (2.10), and x_0, x_1 are given, the next step would be to form the function (2.14). The optimal controls are determined from (2.15) in their dependence on the $x_1, \ldots, x_n; y_0, \ldots, y_n;$ and t. Subsequently, one would then attempt to use $u(t)$ as obtained therefrom to determine the solution of the equations (2.12) and (2.13) subject to the given boundary conditions on x. It is immediately apparent from equation (2.13) that y_0 must be constant, since H is independent of x_0 by definition. Since (2.13) is linear and homogenous in $y_0, y_1, \ldots, y_n$ and since y_0 is constant, there is no loss in generality in choosing $y_0 = -1$ for non-vanishing y_0.

Equation (2.12) for $i = 0$ corresponds to the differentiated form of (2.10), and it is independent of the remaining equations (2.12) and (2.13). Thus, equations (2.12) and (2.13) need to be considered only for $i = 1, \ldots, n$.

2.3   Examples

(1) Time-optimal motion of a particle with bounded acceleration (Pontryagin). Assume that the control function $u(t)$ in the equation

$$\ddot{x} = u \tag{2.16}$$

is subject to the constraint $|u| \leq 1$. For given initial conditions $x(0)$ and $\dot{x}(0)$, that control function is to be found which transfers the system to $x(t_1) = 0, \dot{x}(t_1) = 0$ in as short a time as possible. The cost functional J here is given by

$$J = \int_0^{t_1} dt. \tag{2.17}$$

Equation (2.16) now is written as a first-order system

$$\left.\begin{array}{l} \dot{x}_1 = x_2, \\ \dot{x}_2 = u \end{array}\right\} \tag{2.18}$$

and the function H is defined in accordance with (2.14) as

$$H = y_0 + x_2 y_1 + u y_2. \tag{2.19}$$

Based on (2.13), the adjoint equations are given by

$$\left.\begin{array}{l} \dot{y}_0 = 0, \\ \dot{y}_1 = 0, \\ \dot{y}_2 = -y_1. \end{array}\right\} \tag{2.20}$$

They have the solutions

$$\left.\begin{array}{l} y_0 = C_0, \\ y_1 = C_1, \\ y_2 = C_2 - C_1 t. \end{array}\right\} \tag{2.21}$$

$C_0 \leq 0$ follows from the maximum principle, and one may here choose $C_0 = -1$ without loss of generality. The optimal control is given by

$$\bar{u}(t) = sgn \ y_2(t) = sgn \ (C_2 - C_1 t). \tag{2.22}$$

The optimal control function thus is piecewise constant and assumes only the values $+1$ and -1. Furthermore, it is apparent that the function $\bar{u}(t)$ has at most one point of discontinuity, since the function $C_2 - C_1 t$ clearly has only one zero.

For the time interval for which $u \equiv +1$ holds, equations (2.18) yield

$$\left.\begin{array}{l} x_1 = \frac{1}{2}t^2 + s_2 t + s_1 = \frac{1}{2}(t + s_2)^2 + \left(s_1 - \frac{1}{2}s_2^2\right), \\ x_2 = t + s_2 \end{array}\right\} \tag{2.23}$$

with s_1 and s_2 as constants of integration. The elimination of t yields

$$x_1 = \frac{1}{2}x_2^2 + \left(s_1 - \frac{1}{2}s_2^2\right). \tag{2.24}$$

For $u \equiv +1$ the optimal trajectories thus coincide with the parabolas shown in Fig. 2.3. In an analogous manner, $u \equiv -1$ yields

$$\left.\begin{array}{l} x_1 = -\frac{1}{2}t^2 + \bar{s}_2 t + \bar{s}_1 = -\frac{1}{2}(t - \bar{s}_2)^2 + \left(\bar{s}_1 + \frac{1}{2}\bar{s}_2^{-2}\right) \\ x_2 = -t + \bar{s}_2 \end{array}\right\} \tag{2.25}$$

and

$$x_1 = -\frac{1}{2}x_2^2 + \left(\bar{s}_1 + \frac{1}{2}\bar{s}_2^2\right). \tag{2.26}$$

The corresponding optimal trajectories are shown in Fig. 2.4.

It has been shown already that the optimal control takes on only the values ± 1 and exhibits only one switching point. If $\bar{u}(t)$ first takes on the value $+1$ and then the value -1, then the optimal trajectory must be of the type indicated in Fig. 2.5.

Conversely, if $\bar{u}(t)$ is first equal to -1 and then $+1$, then the optimal trajectory must be of the type indicated in Fig. 2.5. The totality of optimal trajectories has been represented in Fig. 2.6. The heavy curve AOB is called the switching curve for the control function; its upper half OB is given by $x_1 = x_2^2/2$ and its lower half OA by $x_1 = x_2/2^2$. The maximum principle supplies necessary conditions for optimality. Thus, only the trajectories depicted in Fig. 2.6 qualify as candidates for optimal trajectories. Only one such trajectory passes

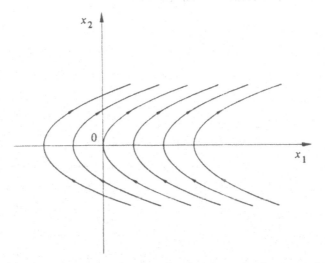

Figure 2.3: The family of optimal trajectories for $u \equiv +1$

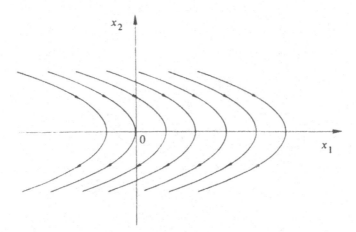

Figure 2.4: The family of optimal trajectories for $u \equiv -1$

through every initial point. If it is known that there exists an optimal control for given initial conditions - as can indeed be shown in the present case - then the field of trajectories shown in Fig. 2.6 actually is that of the optimal trajectories.

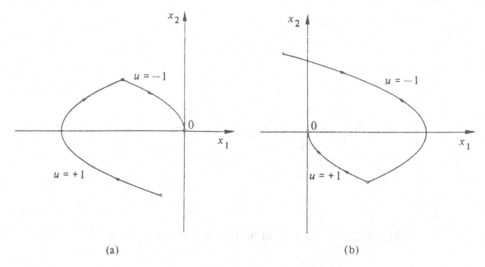

Figure 2.5: Typical optimal trajectories

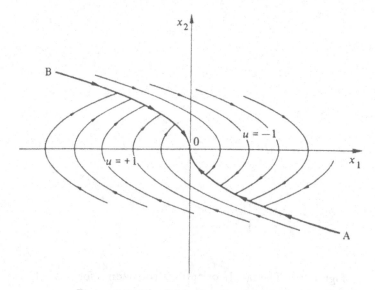

Figure 2.6: The field of optimal trajectories

The solution of the control problem as presented here, may also be interpreted in a somewhat different manner. Let a function $v(x_1, x_2)$ be defined by

$$v(x_1, x_2) = \begin{cases} +1 \text{ for points below the curve AOB and on AO,} \\ -1 \text{ for points above the curve AOB and on OB.} \end{cases}$$

Then, the optimal control may be written in the form $\bar{u}(t) = v(x_1(t), x_2(t))$. When given in terms of the state variables the control function $v(x_1, x_2)$ is called a "closed loop control function", as opposed to $u(t)$ ("open loop control function"). Correspondingly, the differential equations (2.18) may be replaced by $\dot{x}_1 = x_2$ and $\dot{x}_2 = v(x_1, x_2)$.

(2) *Active time-optimal control of an oscillator.* Consider the differential equation

$$\ddot{x} + x = u, \tag{2.27}$$

and write it as the first-order system

$$\left. \begin{aligned} \dot{x}_1 &= x_2, \\ \dot{x}_2 &= -x_1 + u. \end{aligned} \right\} \tag{2.28}$$

Assume that the control function satisfies $|u| \leq 1$. For given initial conditions $u(t)$ again is to be determined in such a way that the system is transferred to $x_1 = 0, x_2 = 0$ in as short a time as possible. With $y_0 = -1$ the function H thus is given by

$$H = -1 + y_1 x_2 - y_2 x_1 + y_2 u, \tag{2.29}$$

and the adjoint equations are obtained as

$$\left. \begin{aligned} \dot{y}_1 &= y_2, \\ \dot{y}_2 &= -y_1. \end{aligned} \right\} \tag{2.30}$$

The solution of equations (2.30) is

$$\left. \begin{aligned} y_1(t) &= -A \cos(t - \alpha_0), \\ y_2(t) &= A \sin(t - \alpha_0), \end{aligned} \right\} \tag{2.31}$$

where $A > 0$ and α_0 are constants of integration. Based on the maximum principle, one obtains

$$\bar{u} = \operatorname{sgn} y_2 = \operatorname{sgn}[A \sin(t - \alpha_0)] = \operatorname{sgn}[\sin(t - \alpha_0)], \tag{2.32}$$

so that $\bar{u}(t)$ is a piecewise constant function with period 2π.

For $\bar{u} \equiv +1$, the general solution of equations (2.28) is given by

$$\left. \begin{aligned} x_1 &= B \sin(t - \beta_0) + 1, \\ x_2 &= B \cos(t - \beta_0), \end{aligned} \right\} \tag{2.33}$$

where B and β_0 are constants of integration. The elimination of t yields the first integral

$$(x_1 - 1)^2 + x_2^2 = B^2. \tag{2.34}$$

In an analogous manner, $\overline{u} \equiv -1$ yields

$$\left. \begin{array}{l} x_1 = \overline{B} \sin\left(t - \overline{\beta}_0\right) - 1, \\ x_2 = \overline{B} \cos\left(t - \overline{\beta}_0\right) \end{array} \right\} \tag{2.35}$$

and

$$(x_1 + 1)^2 + x_2^2 = \overline{B}^2. \tag{2.36}$$

The phase trajectories for (2.28) thus consist of circular arcs with centers at $(1,0)$ and $(-1,0)$, respectively. Since the optimal control function $\overline{u}(t)$ is piecewise constant and takes on only the values $+1$ and -1, the optimal trajectories are composed only of a sequence of such circular arcs. These trajectories can now easily be constructed as shown in Fig. 2.7.

Here also the control problem may again be written in "closed-loop" form with $v(x_1, x_2)$:

$$\left. \begin{array}{l} \dot{x}_1 = x_2, \\ \dot{x}_2 = -x_1 + v(x_1, x_2) \end{array} \right\} \tag{2.37}$$

with a corresponding definition of the optimal control function $v(x_1, x_2)$. When the control function $\overline{u}(t)$ takes on only extreme values and jumps back and forth between them, as was the case in Examples 1 and 2, it is generally referred to as "bang-bang" control.

(3) *Minimum time paths through a region with position-dependent velocity vector.*
This example has no direct application to active damping, but is quite instructive. Consider a ship navigating in the x_1x_2-plane and assume that the motion occurs in a region where a strong current is active. The current has the velocity component $u(x_1, x_2)$ in the direction x_1 and $v(x_1, x_2)$ in the direction of the x_2-axis; the magnitude of the velocity of the ship relative to the water is given by w and the direction of the relative velocity is given by the angle φ of the velocity vector with respect to the x_1-axis. The angle here is the control which is to be chosen in such a way as to minimize the time of travel from a given point A to another given point B (Fig. 2.8). One then has the equations

$$\left. \begin{array}{l} \dot{x}_1 = w \cos \varphi + u(x_1, x_2), \\ \dot{x}_2 = w \sin \varphi + v(x_1, x_2), \end{array} \right\} \tag{2.38}$$

where u, v are known functions of x_1, x_2 and where φ is the control function yet to be determined. With $y_0 = -1$ the Hamiltonian has the form

$$H = -1 + y_1(w \cos \varphi + u) + y_2(w \sin \varphi + v). \tag{2.39}$$

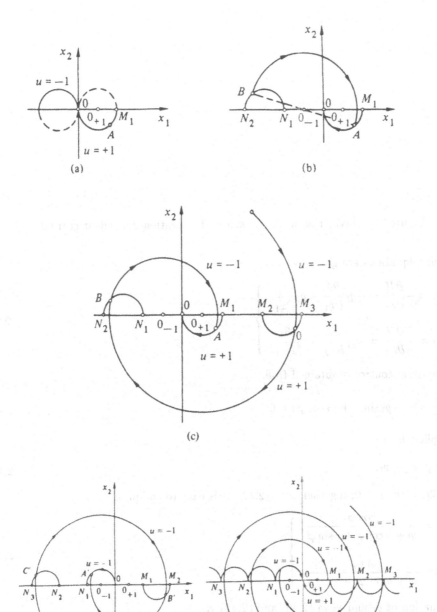

Figure 2.7: Construction of the optimal trajectories

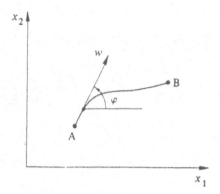

Figure 2.8: Navigation in the presence of a position-dependent current

The adjoint equations are given by

$$\left.\begin{aligned}
\dot{y}_1 &= -\frac{\partial H}{\partial x_1} = -y_1 \frac{\partial u}{\partial x_1} - y_2 \frac{\partial v}{\partial x_1}, \\
\dot{y}_2 &= -\frac{\partial H}{\partial x_2} = -y_1 \frac{\partial u}{\partial x_2} - y_2 \frac{\partial v}{\partial x_2},
\end{aligned}\right\} \tag{2.40}$$

and the optimal control is obtained from

$$\frac{\partial H}{\partial \varphi} = w(-y_1 \sin \varphi + y_2 \cos \varphi) = 0 \tag{2.41}$$

in the explicit form

$$\tan \varphi = y_2/y_1. \tag{2.42}$$

From (2.39) with $H = 0$, together with (2.42), it is easy to compute

$$\left.\begin{aligned}
y_1 &= \frac{\cos \varphi}{w + u \cos \varphi + v \sin \varphi}, \\
y_2 &= \frac{\sin \varphi}{w + u \cos \varphi + v \sin \varphi}.
\end{aligned}\right\} \tag{2.43}$$

The elimination of y_1 and y_2 in (2.40) and (2.42) yields

$$\dot{\varphi} = \sin^2 \varphi \frac{\partial v}{\partial x_1} + \sin \varphi \cos \varphi \left(\frac{\partial u}{\partial x_1} - \frac{\partial v}{\partial x_2}\right) - \cos^2 \varphi \frac{\partial u}{\partial x_2}. \tag{2.44}$$

Finally, the simultaneous solution of (2.38) and (2.44) then results in the optimal trajectories.

An important special case is that in which u and v depend only on the coordinate x_2. The first of equations (2.40) then results in $\dot{y}_1 = $ const., and forming the first of equations (2.43) one obtains

$$\frac{\cos \varphi}{w + u(x_2) \cos \varphi + v(x_2) \sin \varphi} = const. \tag{2.45}$$

Thus, in this case, it is possible to determine the control function φ implicitly as a function of the coordinates.

In the particular case

$$
\begin{aligned}
u(x_1, x_2) &= -w \frac{x_2}{h} \\
v(x_1, x_2) &\equiv 0,
\end{aligned}
\tag{2.46}
$$

with $h = const.$ equation (2.45) yields

$$\frac{\cos \varphi}{w - w \frac{x_2}{h} \cos \varphi} = const. = \frac{\cos \varphi_e}{w}, \tag{2.47}$$

where φ_e is the value of the control at the end-point, that is, at the point $x_1 = 0, x_2 = 0$. It is advantageous to use the angle φ instead of the time as the independent variable in the equations of motion. The function $x_2(\varphi)$ is already known:

$$\frac{x_2}{h} = \frac{1}{\cos \varphi} - \frac{1}{\cos \varphi_e}. \tag{2.48}$$

Based on equation (2.44), one has

$$\dot{\varphi} = \frac{w}{h} \cos^2 \varphi, \tag{2.49}$$

or, equivalently,

$$\frac{dt}{d\varphi} = \frac{h}{w} \frac{1}{\cos^2 \varphi}, \tag{2.50}$$

which results in

$$\frac{w}{h}(t_e - t) \tan \varphi - \tan \varphi_e. \tag{2.51}$$

This relationship may now be used to write the second equation of motion in the form

$$\frac{dx_1}{d\varphi} = -\frac{h}{\cos^3 \varphi}(\cos^2 \varphi + \cos \varphi \cos \varphi_e - 1) \tag{2.52}$$

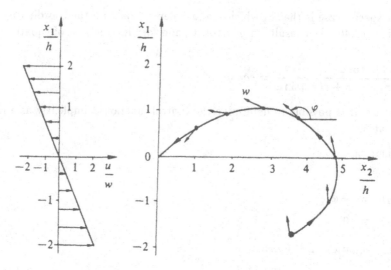

Figure 2.9: a) Flow field,   b) Optimal trajectory

with solution

$$x_1 = \tfrac{h}{2}\left\{sec\ \varphi_e(\tan\varphi_e - \tan\varphi) - \right.$$

$$\left. - \tan\varphi(sec\ \varphi_e - sec\ \varphi) + \ln\frac{\tan\varphi_e + sec\ \varphi_e}{\tan\varphi + sec\ \varphi}\right\}.$$

If one introduces the initial conditions $x_1 = 3.66h, x_2 = -1.86h$ and $\varphi = \varphi_0$, then the expressions for $x_1(\varphi)$ and $x_2(\varphi)$ yield algebraic equations for the two unknowns φ_0 and φ_e. This approach eventually results in $\varphi_0 = 150°$ and $\varphi_e = 240°$. The corresponding optimal trajectory is represented in Fig. 2.9.

References:

[1] **Bryson A.E., Ho Yu-Chi:** Applied optimal control, Ginn, Waltham, Massachussetts, 1969

[2] **Hagedorn P.:** Nonlinear Oscillations, Oxford University Press, Oxford, 1990

[3] **Lee E.B., Markus L.:** Foundations of optimal control theory, John Wiley, New York, 1967

3 Damping in N-degree-of-freedom Systems

3.1 Modal and Non-modal Damping

The linear equations of motion of a forced system with n degrees of freedom are given by

$$\mathbf{M}\ddot{\mathbf{q}} + \mathbf{D}\dot{\mathbf{q}} + \mathbf{C}\mathbf{q} = \mathbf{f}(t) \tag{3.1}$$

with $\mathbf{q} = (q_1, q_2, \ldots, q_n)^T$ and where $\mathbf{M}^T = \mathbf{M} > 0$ is the mass matrix, $\mathbf{C}^T = \mathbf{C} > 0$, the stiffness matrix and $\mathbf{D}^T = \mathbf{D} \geq 0$ the damping matrix. The function $\mathbf{f}(t)$ on the right-hand side is the forcing function or control function. In the undamped case and with the forcing function $f(t) = 0$ (3.1) leads to

$$(-\omega^2 \mathbf{M} + \mathbf{C})\mathbf{l} = 0 \tag{3.2}$$

where all the eigenpairs $\omega_i, \mathbf{l}_i, \; i = 1, 2, \ldots, n$ are real under the assumptions previously made. In (3.2) ω_i are the circular eigenfrequencies and $\mathbf{l}_i$ the eigenvectors. It is well known that the eigenvectors (normalized according to $\mathbf{l}_i^T \mathbf{M} \mathbf{l}_i = 1$) satisfy the orthogonality relations

$$\mathbf{l}_i^T \mathbf{M} \mathbf{l}_j = \delta_{ij}, \quad \mathbf{l}_i^T \mathbf{C} \mathbf{l}_j = \delta_{ij}\omega_i^2 \tag{3.3}$$

if $\omega_i \neq \omega_j$.

Let us now consider the equation

$$\mathbf{M}\ddot{\mathbf{q}} + \mathbf{D}\dot{\mathbf{q}} + \mathbf{C}\mathbf{q} = 0 \tag{3.4}$$

describing the free vibrations of a damped system. The *Ansatz* $\mathbf{q} = \mathbf{l}e^{st}$ now again leads to an eigenvalue problem where, however, both the eigenvalues $\underline{s}_i$ as well as the eigenvectors $\underline{\mathbf{l}}_i$ are complex in general. The eigenpairs $\underline{s}_i, \underline{\mathbf{l}}_i$ appear in complex conjugate pairs. Of course the general solution of (3.4) is obtained by forming a linear combination of the different complex exponential solutions. Dividing respectively the eigenvectors and the eigenvalues into their real and imaginary parts according to

$$\underline{s}_i = -\delta_i + j\omega_{di}, \quad \underline{\mathbf{l}} = \mathbf{k}_i + j\mathbf{v}_i \tag{3.5}$$

leads to the general solution

$$\mathbf{q}(t) = \sum_{i=1}^{n} a_i e^{-\delta_i t} [\mathbf{k}_i \cos(\omega_{di}t + \alpha_i) - \mathbf{v}_i \sin(\omega_{di}t + \alpha_i)] \tag{3.6}$$

where $a_i, \alpha_i, \; i = 1, 2, \ldots, n$ are real integration constants. It can be seen that due to the complexity of the eigenvectors there are normally no time instants t for which all the different generalized coordinates vanish simultaneously if the system oscillates in a complex mode.

It was already examined in lecture 2 that all the real parts δ_i in (3.5) are strictly positive if $\mathbf{D} > 0$. This is a case of complete damping. We may, however, have all δ_i positive even with $\mathbf{D} \geq 0$. This case of *pervasive damping* is illustrated in the example of Fig. 3.1.

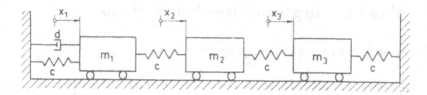

Figure 3.1: Illustration of pervasive damping

The system matrices of the system of Fig. 3.1 are given by

$$\mathbf{M} = \begin{pmatrix} m_1 & 0 & 0 \\ 0 & m_2 & 0 \\ 0 & 0 & m_3 \end{pmatrix}, \quad \mathbf{C} = \begin{pmatrix} 2c & -c & 0 \\ -c & 2c & -c \\ 0 & -c & 2c \end{pmatrix}, \quad \mathbf{D} = \begin{pmatrix} d & 0 & 0 \\ 0 & 0 & 0 \\ 0 & 0 & 0 \end{pmatrix}. \tag{3.7}$$

It is a simple exercise to show that the system represented in Fig. 3.1 does not exhibit undamped motions, i.e. that $\delta_i > 0, i = 1, 2, 3$.

In the previous lecture a condition for pervasiveness of the damping matrix had been formulated. Here we will briefly reconsider this problem. Suppose that (3.4) admits an undamped motion, which can only be a harmonic solution. Then

$$(-\omega^2 \mathbf{M} + \mathbf{C})\mathbf{l} = \mathbf{0}, \quad \mathbf{D}\mathbf{l} = \mathbf{0} \tag{3.8}$$

hold simultaneously. These equations only have a non-trivial solution in l if

$$\mathrm{rank}\,(-\omega^2 \mathbf{M} + \mathbf{C}, \mathbf{D}) < n \tag{3.9}$$

holds. For

$$\mathrm{rank}\,(-\omega^2 \mathbf{M} + \mathbf{C}, \mathbf{D}) = n \tag{3.10}$$

the system (3.8) only admits the trivial solution in l. Both the conditions (3.9) and (3.10) need to be verified for the eigenfrequencies ω of the undamped system only. Equation (3.10) is a reformulation of the condition for the pervasiveness of damping which was already given in lecture 2.

Usually the eigenvectors of the damped system are different from the eigenvectors of the undamped system. There are, however, certain cases in which the (real) eigenvectors of the undamped system are identical to the ones of the damped system. We will briefly consider the conditions on $\mathbf{D}$ for this to occur. From the orthogonality conditions (3.3) it is clear that if $\mathbf{D}$ is proportional to $\mathbf{M}$, then the real eigenvectors of the undamped system are orthogonal also with respect to $\mathbf{D}$, which does then not affect the eigenvectors.

The same holds for $\mathbf{D}$ proportional to $\mathbf{C}$ and also for

$$\mathbf{D} = \alpha\mathbf{M} + \beta\mathbf{C}. \tag{3.11}$$

This is a sufficient condition for the coincidence of the eigenvectors of the damped and undamped systems, it is, however, not a necessary condition. Even for more general matrices $\mathbf{D}$ the coincidence of the eigenvectors is possible. Let us derive this more general condition. The eigenvalue problem

$$(\underline{s}^2\mathbf{M} + \underline{s}\mathbf{D} + \mathbf{C})\mathbf{l} = 0 \tag{3.12}$$

can be transformed into

$$(\underline{s}^2\mathbf{I} + \underline{s}\mathbf{M}^{-1}\mathbf{D} + \mathbf{M}^{-1}\mathbf{C})\mathbf{l} = 0 \tag{3.13}$$

by multiplying it by $\mathbf{M}^{-1}$. Multiplying again by $\mathbf{M}^{-1}\mathbf{C}$ one has

$$(\underline{s}^2\mathbf{M}^{-1}\mathbf{C} + \underline{s}(\mathbf{M}^{-1}\mathbf{C})(\mathbf{M}^{-1}\mathbf{D}) + (\mathbf{M}^{-1}\mathbf{C})^2)\mathbf{l} = 0. \tag{3.14}$$

If, and only if the commutation condition

$$(\mathbf{M}^{-1}\mathbf{C})(\mathbf{M}^{-1}\mathbf{D}) = (\mathbf{M}^{-1}\mathbf{D})(\mathbf{M}^{-1}\mathbf{C}) \tag{3.15}$$

holds, this can be written as

$$(\underline{s}^2\mathbf{I} + \underline{s}\mathbf{M}^{-1}\mathbf{D} + \mathbf{M}^{-1}\mathbf{C})(\mathbf{M}^{-1}\mathbf{C})\mathbf{l} = 0 \tag{3.16}$$

or

$$(\underline{s}^2\mathbf{M} + \underline{s}\mathbf{D} + \mathbf{C})\mathbf{M}^{-1}\mathbf{C}\,\mathbf{l} = 0. \tag{3.17}$$

We now compare (3.12) to (3.17) and see that if $(\underline{s}, \mathbf{l})$ is an eigenpair then also $(\underline{s}, \mathbf{M}^{-1}\mathbf{Cl})$ is an eigenpair of the same eigenvalue problem. Suppose that $\underline{s}$ is simple, then only one eigenvector corresponds to the eigenvalues, and we have

$$\mathbf{M}^{-1}\mathbf{Cl} = \mu\mathbf{l} \tag{3.18}$$

which can also be written as

$$\mathbf{Cl} = \mu\mathbf{Ml} \tag{3.19}$$

or

$$(\mathbf{C} - \mu\mathbf{M})\mathbf{l} = 0. \tag{3.20}$$

Therefore the eigenvector $\mathbf{l}$ of the damped system is also a (real) eigenvector of the undamped system. The condition (3.15) is necessary and sufficient for having *modal damping* (i.e. damping that does not couple the eigenmodes of the undamped system).

If the damping is modal then a coordinate transformation using the modal matrix transforms the homogenous system into

$$\ddot{p}_k + 2D_k\omega_k\dot{p}_k + \omega_k^2 p_k = 0 \qquad (3.21)$$

where $k = 1, 2, \ldots, n$. Only in the case of modal damping the n-degree-of-freedom system can be uncoupled into n damped linear oscillators with a real coordinate transformation.

If the damping matrix is not precisely known, then frequently the eigenvalue problem is solved for the undamped system for which the normal coordinates and the eigenfrequencies are determined. The damping ratios D_k are then assumed a posteriori. Building codes often specify the damping ratios to be assumed in the calculation. For example DIN 4149 recommends to use approximately $D = 0.051$ for all modes in the study of earthquake vibrations of nuclear power stations. In steel constructions on the other hand very low values such as for example $D = 0.0017$ are often specified.

3.2   The Case of Small Damping - A Perturbational Approach

It has already be seen that unless some artificial damping is introduced in a structure the damping terms are usually small. In what follows we will assume that the free vibrations of a damped system are described by

$$\mathbf{M\ddot{q}} + \epsilon\mathbf{D\dot{q}} + \mathbf{Cq} = 0 \qquad (3.22)$$

with $\epsilon \ll 1$. We will give a simple perturbational approach for the solution of the corresponding eigenvalue problem as a function of ϵ. The *Ansatz*

$$\mathbf{q}(t) = \mathbf{l}e^{st} \qquad (3.23)$$

leads to

$$(\underline{s}^2\mathbf{M} + \epsilon\underline{s}\mathbf{D} + \mathbf{C})\underline{\mathbf{l}} = 0. \qquad (3.24)$$

We assume both the complex eigenvalue $\underline{s}$ as well as the complex eigenvector $\underline{\mathbf{l}}$ as being represented by a power series in ϵ according to

$$\underline{s} = \underline{s}_0 + \epsilon\underline{s}_1 + \epsilon^2\underline{s}_2 + \ldots \qquad (3.25)$$

$$\underline{\mathbf{l}} = \underline{\mathbf{l}}_0 + \epsilon\underline{\mathbf{l}}_1 + \epsilon^2\underline{\mathbf{l}}_2 + \ldots. \qquad (3.26)$$

It is well known that such an expansion needs not to converge in general. For $\epsilon = 0$ we obtain the solution of the undamped problem, namely

$$\underline{s}_0 = j\omega_k \qquad (3.27)$$

$$\underline{\mathbf{l}}_0 = \mathbf{r}_k. \qquad (3.28)$$

Here r_k is the k-th eigenvector of the undamped system normalized with respect to the mass matrix. In what follows we will focus our attention on the first mode only, so that we omit the index k which should be used to distinguish the different eigenpairs in (3.25), (3.26). The eigenvector $\underline{l}$ of the perturbed problem will be normalized according to

$$\underline{l}^T M r_1 = 1. \tag{3.29}$$

Therefore also

$$\underline{l}_1^T M r_1 = 0, \quad \underline{l}_2^T M r_1 = 0, \quad \underline{l}_3^T M r_1 = 0, \ldots \tag{3.30}$$

holds. Substituting (3.25) and (3.26) into (3.24) and comparing terms of the same order of magnitude gives

$$\epsilon^0 : (-\omega_1^2 M + C) r_1 = 0, \tag{3.31}$$

$$\epsilon^1 : (-\omega_1^2 M + C)\underline{l}_1 + (2j\omega_1 \underline{s}_1 M + j\omega_1 D) r_1 = 0. \tag{3.32}$$

$$\epsilon^2 : (-\omega_1^2 M + C)\underline{l}_2 + (2j\omega_1 \underline{s}_1 M + j\omega_1 D)\underline{l}_1 + $$
$$+ \left[(\underline{s}_1^2 + 2j\omega_1 \underline{s}_2) M + \underline{s}_1 D \right] r_1 = 0. \tag{3.33}$$

$\underline{l}_1$ being orthogonal to r_1 with respect to M, it can be written as

$$\underline{l}_1 = \underline{\beta}_{1,2} r_2 + \underline{\beta}_{1,3} r_3 + \ldots + \underline{\beta}_{1,n} r_n \tag{3.34}$$

where the coefficients $\underline{\beta}_{1,2}, \ldots, \underline{\beta}_{1,n}$ have to be determined. Multiplying (3.32) by r_1^T from the left gives

$$\underline{s}_1 = -\frac{1}{2} r_1^T D r_1. \tag{3.35}$$

The first correction to the eigenvalue of the undamped system has therefore been obtained. It is always real and non-positive. Similarly, multiplying (3.32) from the left by r_u^T gives an equation for the determination of $\underline{\beta}_{1,u}$. Together with (3.34) this gives the first correction to the eigenvector

$$\underline{l}_1 = j\omega_1 \sum_{u=2}^{n} \frac{1}{\omega_1^2 - \omega_u^2} (r_u^T D r_1) r_u. \tag{3.36}$$

It can be seen that this correction is always purely imaginary. Similarly higher-order approximations to the eigenvalues and eigenvectors can easily be computed. The case of multiple eigenvalues has to be considered separately, as is obvious already from equation (3.36).

The approximate solution of the eigenvalue problem of the damped system can be useful in some practical cases. It is then usually sufficient to use the first-order approximation.

3.3   Mechanical Impedance, Vibration Absorbers, Impedance Matching

We reconsider equation (3.1) for the case of harmonic excitation. In complex notation we have

$$\mathbf{M\ddot{q}} + \mathbf{D\dot{q}} + \mathbf{Cq} = \mathbf{\hat{f}}e^{j\Omega t}. \tag{3.37}$$

We are interested in the steady-state solution

$$\mathbf{q}(t) = \mathbf{\hat{q}}e^{j\Omega t}. \tag{3.38}$$

Substituting into (3.37) gives

$$\mathbf{\hat{q}} = \mathbf{\underline{G}}(\Omega)\mathbf{\hat{f}} \tag{3.39}$$

with

$$\mathbf{\underline{G}}(\Omega) = (-\Omega^2\mathbf{M} + j\Omega\mathbf{D} + \mathbf{C})^{-1}. \tag{3.40}$$

Equation (3.40) defines the complex compliance matrix of an n-degree-of-freedom system. Often we are more interested in the complex impedance matrix, using the velocities as inputs and forces as outputs. The complex impedance matrix is therefore given by

$$\mathbf{\underline{Z}}(\Omega) = \frac{1}{j\Omega}(-\Omega^2\mathbf{M} + j\Omega\mathbf{D} + \mathbf{C}). \tag{3.41}$$

If in equation (3.37) only the first m force components are different from zero and $\underline{f}_{m+1}$, $\underline{f}_{m+2}, \ldots, \underline{f}_n$ are equal to zero then

$$\left[\mathbf{\hat{q}}\right]_m = [\mathbf{\underline{G}}(\Omega)]_m \left[\mathbf{\hat{f}}\right]_m \tag{3.42}$$

holds with

$$\left[\mathbf{\hat{q}}\right]_m = \left(\hat{q}_1, \hat{q}_2, \ldots, \hat{q}_m\right)^T, \tag{3.43}$$

$$\left[\mathbf{\hat{f}}\right]_m = \left(\hat{f}_1, \hat{f}_2, \ldots, \hat{f}_m\right)^T. \tag{3.44}$$

The complex compliance matrix and the complex impedance matrix of the reduced system of m inputs and m outputs are respectively given by

$$[\mathbf{\underline{G}}(\Omega)]_m = \left[(-\Omega^2\mathbf{M} + j\Omega\mathbf{D} + \mathbf{C})^{-1}\right]_m \tag{3.45}$$

$$\mathbf{\underline{Z}}^{(m)}(\Omega) = \frac{1}{j\Omega}\left[(-\Omega^2\mathbf{M} + j\Omega\mathbf{D} + \mathbf{C})^{-1}\right]_m^{-1}. \tag{3.46}$$

Note that in equation (3.45), (3.46) the symbol $[\ ]_m$ means that on the right-hand side the upper $m \times m$ sub-matrix is to be selected. However, in (3.45), (3.46) first the $n \times n$ matrices have to be inverted.

Complex impedance matrices of this type are often measured, particulary in experimental modal analysis. The complex impedances completely represent the linear dynamic behaviour of the system at its point of interconnection to other systems. Also dynamic vibration absorbers in the steady-state case can be completely described by their impedance matrix.

Consider the forced harmonic vibrations of the system shown in Fig. 3.2a. If the forcing frequency Ω is known, these vibrations can be conveniently limited by means of a vibration absorber as shown in Fig. 3.2b. The vibration absorber depicted in Fig. 3.2 has been studied extensively by Den Hartog. In what follows we briefly discuss the somewhat more general problem illustrated by Fig. 3.3. The vibrations of the base block are described by

$$m_1 \ddot{x}_1 + d_1 \dot{x}_1 + c_1 x_1 = \hat{f}_1 \cos \Omega t + p(t) \tag{3.47}$$

where $p(t)$ is the force acting on the base due to the vibration absorber. The vibration absorber is characterized completely by its complex impedance which may, for example, be given in a catalogue from a manufacturer. The number of degrees of freedom and the particular design of the impedance is irrelevant at this point.

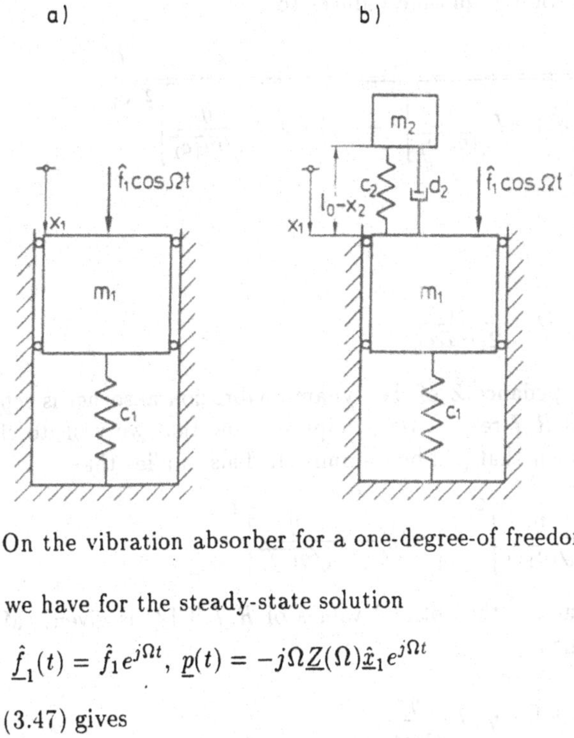

Figure 3.2: On the vibration absorber for a one-degree-of freedom system

In complex notation we have for the steady-state solution

$$\underline{x}_1(t) = \hat{\underline{x}}_1 e^{j\Omega t}, \quad \underline{f}_1(t) = \hat{f}_1 e^{j\Omega t}, \quad \underline{p}(t) = -j\Omega \underline{Z}(\Omega)\hat{\underline{x}}_1 e^{j\Omega t} \tag{3.48}$$

which together with (3.47) gives

$$\left[-m_1 \Omega^2 + (d_1 + \underline{Z}(\Omega))j\Omega + c_1 \right] \hat{\underline{x}}_1 = \hat{\underline{f}}_1. \tag{3.49}$$

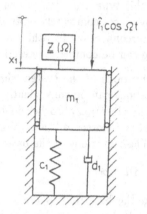

Figure 3.3: Fundamental system with vibration absorber of impedance $\underline{Z}(\Omega)$

Solving for the vibration amplitude reduces to

$$\hat{\underline{x}}_1 = \frac{1}{\sqrt{\left[(1 - \eta^2) - I\dfrac{\eta}{\sqrt{m_1 c_1}}\right]^2 + \left[2D\eta + R\dfrac{\eta}{\sqrt{m_1 c_1}}\right]^2}} \frac{\hat{f}_1}{c_1} \tag{3.50}$$

with

$$\underline{Z} = R + jI, \tag{3.51}$$

$$\eta = \frac{\Omega}{\sqrt{c_1/m_1}}, \quad D = \frac{d_1}{2\sqrt{m_1 c_1}}. \tag{3.52}$$

Here the complex impedance $\underline{Z}$ of the dynamic vibration absorber is represented by its real and imaginary parts R, I respectively. Suppose now that we wish to choose an "optimal" vibration absorber such that $|\hat{\underline{x}}_1|$ be minimized. This implies that

$$\left[(1 - \eta^2) - I\frac{\eta}{\sqrt{m_1 c_1}}\right]^2 + \left[2D\eta + R\frac{\eta}{\sqrt{m_1 c_1}}\right]^2 \tag{3.53}$$

be maximized. What are the optimal values of R, I if $|\underline{Z}|$ is given? Maximizing (3.53) is equivalent to maximizing

$$4DR\frac{\eta^2}{\sqrt{m_1 c_1}} - 2(1 - \eta^2)I\frac{\eta}{\sqrt{m_1 c_1}} \tag{3.54}$$

with respect to R, I. The solution is that R, I have to be chosen such that

$$\frac{R}{I} = \frac{2D\eta}{-(1 - \eta^2)}, \quad R > 0 \tag{3.55}$$

holds. This can easily be seen by regarding (3.54) as the scalar product of the two vectors

$$(R, I) \quad \text{and} \quad (4D\eta^2/\sqrt{m_1 c_1}, \ -2\eta(1 - \eta^2)/\sqrt{m_1 c_1}). \tag{3.56}$$

The scalar product assumes its maximum value if the two vectors are parallel and this gives (3.55). The optimal impedance of the dynamic vibration absorber shown in Fig.3.3, subject to the condition of a constant absolute value of the impedance, has therefore been solved. It is not surprising that it satisfies a condition which is very closely related to the well-known impedance matching. The effective value of the power dissipated in the dynamic absorber is given by

$$\overline{P} = \frac{1}{2}R\dot{x}_1^2. \tag{3.57}$$

3.4 The Elastically Supported Rigid Body, Principal Elastic Axes

The system shown in Fig. 3.4 corresponds to a three d.o.f. system excited by an unbalanced rotor. If a system of dynamic vibration absorbers is to be designed for this case, it will in general have to be composed of three vibration absorbers of the Den Hartog type, tuned to the frequency Ω.

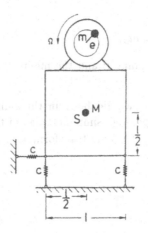

Figure 3.4: Elastically supported planar base block with rotating machine

In three-dimensional space a rigid body has six degrees of freedom so that six of these vibration absorbers would be needed in general. However, usually some of the modes of this six-degree-of-freedom system are more important than others, and vibration absorbers can be designed in such a way as to interact with the most important mode. This may considerably simplify the design of vibration absorbers for six d.o.f. systems. It requires, however, that the modes are known.

In three-dimensional space small displacements of a rigid body can be characterized by

$$\mathbf{q}(t) = (x_1, x_2, x_3, \varphi_1, \varphi_2, \varphi_3)^T \tag{3.58}$$

where

$$\vec{s} = (x_1, x_2, x_3)^T, \quad \vec{\varphi} = (\varphi_1, \varphi_2, \varphi_3)^T \tag{3.59}$$

contain the cartesian coordinates of the linear displacement $\vec{s} = \overrightarrow{PP'}$ of a point P of the body from its old to its new position P' (translation) and $\vec{\varphi}$ is the rotation vector about P. The potential energy is

$$U = \frac{1}{2}\mathbf{q}^T \mathbf{C}\mathbf{q} \tag{3.60}$$

with $\mathbf{C}^T = \mathbf{C} > 0$. This leads to

$$\mathbf{Q} = \mathbf{C}\mathbf{q}, \tag{3.61}$$

where

$$\mathbf{Q} = (f_1, f_2, f_3, m_1, m_2, m_3). \tag{3.62}$$

Here

$$\vec{f} = (f_1, f_2, f_3)^T, \quad \vec{m} = (m_1, m_2, m_3)^T \tag{3.63}$$

are respectively the force applied at point P and the moment (couple) to produce the displacement $\mathbf{q}$.

For a general linear elastic support of a rigid body in three-dimensional space the matrix $\mathbf{C}$ is full. Let us briefly look at the possible simplifications of the matrix $\mathbf{C}$. We note that both $\mathbf{q}$ and $\mathbf{Q}$ are transformed using the same transformation matrix $\mathbf{T}$ if we go from one cartesian reference system to another.

$$\mathbf{Q}' = \mathbf{T}\mathbf{Q}, \quad \mathbf{q} = \mathbf{T}^T\mathbf{q}', \tag{3.64}$$

$$\mathbf{C}' = \mathbf{T}\mathbf{C}\mathbf{T}^T. \tag{3.65}$$

$\mathbf{C}'$ is the 6×6 stiffness matrix in the new reference system. A rotation in three-dimensional space from one system of cartesian (orthogonal) axes to another is given by

$$\mathbf{T}_{rot} = \begin{pmatrix} \mathbf{G} & \mathbf{0} \\ \mathbf{0} & \mathbf{G} \end{pmatrix}, \tag{3.66}$$

where $\mathbf{G}$ is the three-dimensional rotation matrix. Similarly a translation is described by

$$
\mathbf{T}_{trans} = \begin{pmatrix}
1 & 0 & 0 & 0 & 0 & 0 \\
0 & 1 & 0 & 0 & 0 & 0 \\
0 & 0 & 1 & 0 & 0 & 0 \\
0 & +z_0 & -y_0 & 1 & 0 & 0 \\
-z_0 & 0 & +x_0 & 0 & 1 & 0 \\
+y_0 & -x_0 & 0 & 0 & 0 & 1
\end{pmatrix} ,
\tag{3.67}
$$

where x_0, y_0, z_0 are the coordinates of the origin of the new reference system given in the old system. A general transformation matrix $\mathbf{T}$ can be written as product of two matrices of the types (3.66) and (3.67). It is an interesting problem to study which coordinate transformations give simple representations of the stiffness matrix. It can be shown that in general three central displacement axes (orthogonal but not intersecting in general), as well as three central rotation axes (orthogonal but not intersecting in general) exist. The central displacement (rotation) axis has the property that a linear displacement in its direction (a rotation about this axis) corresponds only to a force and a moment vector in its direction (i.e. a "wrench"). General elastic supports therefore do not admit axes for which a translation of the rigid body corresponds exclusively to a force and the rotation of the body exclusively to a moment!

Of course, in the linear vibrations of an elastically supported body, also the inertia properties are important. The transformation of the inertia matrix of a rigid body is, however, much more widely known than the transformation of the elastic support which we discussed above. The solution of the linear vibration problem of an elastically supported body leads to six eigenmodes. In each mode the body's motion can be described as a "wrench" motion and the corresponding absorber also has to be of the "wrench" type (i.e. rotation about one axis as well as translation in the direction of the same axis). The knowledge of these modes is obviously important in designing simple vibration absorbers, be they passive or active.

References

[1] **Hagedorn P., Otterbein S.**: Technische Schwingungslehre, Springer, Berlin, 1987

[2] **Schiel F.**: Statik der Pfahlwerke, 2. Auflage, Springer, Berlin, 1987

4   Nonlinearities in Structural Vibrations

4.1   Introduction

Although the majority of the problems of structural dynamics can be described by linear differential equations at least in a first approximation, the significance of nonlinear systems in engineering vibration problems is continuously increasing. This is, on one hand, due to improved computational methods and facilities, on the other hand to the higher requirements with respect to precision and safety. In many cases, however, there is only limited knowledge available on the nonlinear material behaviour, the dynamic behaviour at the system boundaries and the points of interconnection, and also on the loads acting on the real structure; in these cases sophisticated nonlinear modelling may not be adequate before realistic values have been found for these data.

In general, the nonlinear problems of structural vibrations can be described by differential equations of the type

$$\mathbf{M}(\mathbf{q}, t)\ddot{\mathbf{q}} = \mathbf{f}_1(\mathbf{q}, \dot{\mathbf{q}}, t), \tag{4.1}$$

at least after discretization. In this formula $\mathbf{q}$ is the n-dimensional vector of the generalized coordinates, and the $n \times n$ matrix $\mathbf{M}$ is non-singular for all values of $\mathbf{q}$ and t. Equation (4.1) shows already one of the main properties of nonlinear oscillations of mechanical systems: it is in general not possible to solve the equations explicitly and analytically with respect to the highest derivatives (i.e. it is usually not possible to invert the matrix $\mathbf{M}$ symbolically), and this has consequences for the numerical treatment.

The dependence of the mass matrix $\mathbf{M}$ on the generalized coordinates is typically due to *geometric* nonlinearities. This brings us to possible classifications of the nonlinearities in structures as well as of the nonlinear oscillations. Such a classification is neither simple nor unique, due to the many different phenomena and sources of nonlinearities. It is, however, as a rule not difficult to distinguish between geometric nonlinearities and other nonlinearities due to the material properties and to the connections existing between the different parts of a structure. The number of degrees of freedom is a completely different criterion of classification which may, however, be of extreme importance for practical purposes. This is due to the fact that technical problems described by a very large number of degrees of freedom frequently have to be treated in a different manner than systems of few degrees of freedom. In large systems it is usually impossible to oversee the complete variety of different types of solutions, which on the other hand holds true for small systems such as for example the classical VAN DER POL oscillator.

In the modern theory of nonlinear dynamics concepts such as *catastrophe theory, bifurcation theory, chaotic motions, etc.* are of importance. It must, however, be said that these concepts so far have had little impact on technical engineering problems, possibly with the exception of bifurcation aspects, although they can be fascinating from the mathematical point of view. In some cases, however, they permit new and deeper views of known phenomena.

In what follows we will examine an approximate method which turns out to be quite useful in the analysis of simple nonlinear systems.

4.2 Slowly Changing Amplitude and Phase

Consider the equation

$$\ddot{x} + \omega_0^2 x = f(x, \dot{x}), \tag{4.2}$$

where $f(x, \dot{x})$ is an arbitrary (nonlinear) function, and introduce the coordinate transformation from $x, \dot{x}$ to a, ψ as defined by

$$x(t) = a(t) \sin(\omega_0 t + \psi(t)), \tag{4.3}$$

$$\dot{x}(t) = a(t) \omega_0 \cos(\omega_0 t + \psi(t)). \tag{4.4}$$

The differentiation of (4.3) with respect to time and a comparison of the result with the second yield

$$\dot{a}(t) \sin(\omega_0 t + \psi(t)) + a(t) \dot{\psi}(t) \cos(\omega_0 t + \psi(t)) = 0. \tag{4.5}$$

Equation (4.2) can now be written as

$$\dot{a} = \frac{1}{\omega_0} f(a \sin(\omega_0 t + \psi), a\omega_0 \cos(\omega_0 t + \psi)) \cos(\omega_0 t + \psi), \tag{4.6}$$

$$\dot{\psi} = -\frac{1}{\omega_0 a} f(a \sin(\omega_0 t + \psi), a\omega_0 \cos(\omega_0 t + \psi)) \sin(\omega_0 t + \psi). \tag{4.7}$$

Up to this point, no more than a coordinate transformation has been carried out, and the differential equations (4.6), (4.7) are still exactly equivalent to (4.2). If $f(x, \dot{x})$ is "small", then $\dot{a}$ and $\dot{\psi}$ are also small, that is, the amplitude a and the phase ψ change only slowly. "Slowly" here means that the value of ψ in the argument $(\omega_0 t + \psi)$ as well as the value of $a(t)$ remain neraly constant during a time interval of duration $T_0 = 2\pi/\omega_0$. Since it generally is not possible to obtain an exact solution to (4.2), the differential equation is simplified by replacing the right-hand sides by their temporal mean over the interval $[t, t + T_0]$. In forming this mean value, a and ψ will be kept constant on the right-hand side of (4.6), (4.7). Equations (4.6), (4.7) thus are replaced by

$$\dot{a} = \frac{1}{\omega_0 2\pi} \int_0^{2\pi} f(a \sin(\Theta + \psi), a\omega_0 \cos(\Theta + \psi)) \cos(\Theta + \psi) d\Theta, \tag{4.8}$$

$$\dot{\psi} = \frac{-1}{a\omega_0 2\pi} \int_0^{2\pi} f(a \sin(\Theta + \psi), a\omega_0 \cos(\Theta + \psi)) \sin(\Theta + \psi) d\Theta. \tag{4.9}$$

The result is an autonomous (that is, time-independent) system of equations providing a first approximation to the time dependence of the amplitude and phase. For this reason the method is often called the method of slowly changing phase and amplitude.

As a first example we apply this procedure to the case of linear damping and a cubic nonlinearity; with

$$
\begin{aligned}
f(x,\dot{x}) &= \omega_0^2 \frac{x^3}{6} - 2\delta \dot{x} \\
&= \frac{\omega_0^2}{6} a^3 \sin^3(\omega_0 t + \psi) - 2\delta a \omega_0 \cos(\omega_0 t + \psi) \\
&= \frac{\omega_0^2}{24} a^3 [3\sin(\omega_0 t + \psi) - \sin(3\omega_0 + 3\psi)] - \\
&\quad - 2\delta a \omega_0 \cos(\omega_0 t + \psi)
\end{aligned}
\tag{4.10}
$$

equations (4.7), (4.8) result in

$$
\dot{a} = -\delta a,
\tag{4.11}
$$

$$
\dot{\psi} = -\omega_0 \frac{a^2}{16}.
\tag{4.12}
$$

An integration of the first differential equation, a substitution of the result into the second, and a use of the initial conditions $a(0) = C_0$, $\psi(0) = 0$ eventually yield

$$
a(t) = C_0 e^{-\delta t},
\tag{4.13}
$$

$$
\psi(t) = C_0^2 \frac{\omega_0}{32\delta}(e^{-2\delta t} - 1).
\tag{4.14}
$$

In accordance with (4.3) one thus obtains the approximate solution

$$
x(t) = C_0 e^{-\delta t} \sin\left[\omega_0 t + C_0^2 \frac{\omega_0}{32\delta}(e^{-2\delta t} - 1)\right].
\tag{4.15}
$$

In a second example we treat the case of Coulomb damping and a third-order spring, giving

$$
\begin{aligned}
f(x,\dot{x}) &= \omega_0^2 \frac{x^3}{6} - \rho \, \text{sgn}(\dot{x}) \\
&= \frac{\omega_0^2}{24} a^3 \{3\sin(\omega_0 t + \psi) - \sin(3\omega_0 t + 3\psi)\} - \\
&\quad - \rho \, \text{sgn}\{\cos(\omega_0 t + \psi)\}.
\end{aligned}
\tag{4.16}
$$

The use of equations (4.8), (4.9) then results in

$$
\dot{a} = -\frac{2\rho}{\pi \omega_0},
\tag{4.17}
$$

$$
\dot{\psi} = -\omega_0 \frac{a^2}{16}.
\tag{4.18}
$$

Integration yields

$$a(t) = C_0 - \frac{2\rho}{\pi\omega_0}t, \tag{4.19}$$

$$\psi(t) = -C_0^2\frac{\omega_0}{16}t + C_0\frac{\rho}{8\pi}t^2 - \frac{\rho^2}{12\pi^2\omega_0}t^3 \tag{4.20}$$

so that

$$\begin{aligned}
x(t) &= C_0\left(1 - \frac{2\rho}{\pi\omega_0 C_0}t\right)\sin\left\{\omega_0\left(1 - \frac{C_0^2}{16}\right)t +\right. \\
&\quad + \left. C_0\frac{\rho}{8\pi}t^2 - \frac{\rho^2}{12\pi^2\omega_0}t^3\right\}
\end{aligned} \tag{4.21}$$

is the approximate solution.

The quality of the results which one may obtain under the appropriate circumstances may be deduced, for example, from the phase diagrams in Figs. 4.1 and 4.2.

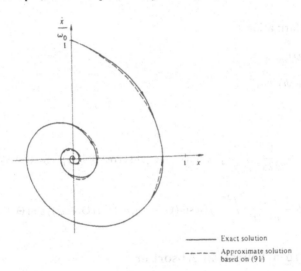

—————— Exact solution

– – – – – Approximate solution
based on (91)

Figure 4.1: Phase diagram of damped linear free oscillations of the mathematical pendulum with $C_0 = 1, \omega_0 = 1, \delta = 0.2$

In the case of forced oscillations we proceed in a similar manner. Suppose that we are interested in solutions of

$$\ddot{x} + \omega_0^2 x - f(x, \dot{x}) = P\sin\Omega t. \tag{4.22}$$

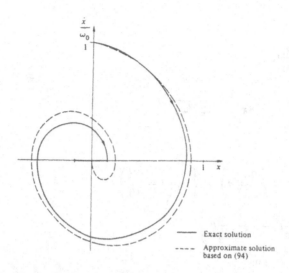

Figure 4.2: Phase diagram of the free oscillation of the mathematical pendulum with Coulomb damping and with $C_0 = 1, \omega_0 = 1, \rho = 0.16$

The coordinate transformation

$$x = a\sin(\Omega t + \psi), \tag{4.23}$$

$$\dot{x} = a\Omega\cos(\Omega t + \psi) \tag{4.24}$$

then yields

$$\dot{a} = -\frac{P}{2\Omega}\sin\psi + \frac{1}{\Omega 2\pi}\int_0^{2\pi} f(a\sin(\Theta + \psi), a\Omega\cos(\Theta + \psi))\cos(\Theta + \psi)d\Theta, \tag{4.25}$$

$$\dot{\psi} = -\frac{P}{2\Omega a}\cos\psi - \frac{1}{a\Omega 2\pi}\int_0^{2\pi} f(a\sin(\Theta + \psi), a\Omega\cos(\Theta + \psi))\sin(\Theta + \psi)d\Theta \tag{4.26}$$

4.3 Application to a Vibration Absorber

We now apply the method to the equations of motion of the liquid-filled damper shown in Fig. 4.3, which has been used experimentally in high-voltage overhead transmission lines for quenching oscillations of very low frequencies ($f \ll 1Hz$). It consists of a liquid-filled cylindric container with an inner box and plates with holes. Its working principle follows from the diagram of Fig. 4.4. In the design of the damper the main problem was to choose the dimensions of the different parts in such a way as to maximize the dissipation of mechanical

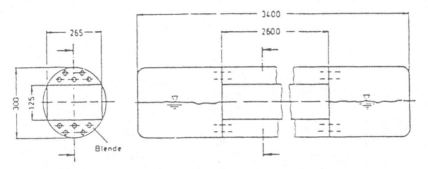

Figure 4.3: Liquid-filled damper

energy during the damper's harmonic rotational motion according to $\beta(t) = \hat{\beta}\sin\Omega t$ (see Fig. 4.4b).

The instationary BERNOULLI equation can be written for this system as

$$\left[H + \frac{l}{\alpha}\right]\ddot{h} + \frac{2\zeta}{\alpha^2}\dot{h}^2\,sgn\dot{h} + 2gh = \left[g - \frac{H}{2}\Omega^2\right]l\hat{\beta}\sin\Omega t. \tag{4.27}$$

In this equation the parameters H and α are given by $H/2 = h_1 + h_2, \alpha = A/A_1$ and ζ is the loss factor which depends on the diaphragm used; all the other symbols can immediately be identified in Fig. 4.4. The changes in the angle β are small so that $\sin\beta \approx \beta$. It can easily be seen that the excitation term on the right-hand side of (4.22) vanishes for

$$\Omega = \Omega_b := \sqrt{\frac{2g}{H}}. \tag{4.28}$$

For this value of the excitation frequency the liquid does not oscillate with respect to the container (in steady state), and the frequency Ω_b corresponds to the eigenfrequency of the liquid column of an U-shaped tube of constant cross-section and length H. In non-dimensional form the equation of motion can be written as

$$\ddot{\overline{h}} + B\dot{\overline{h}}^2\,sgn\,\dot{\overline{h}} + \overline{h} = E\sin\eta\tau. \tag{4.29}$$

The *Ansatz* of the slowly varying face and amplitude according to

$$\overline{h}(\tau) = a(\tau)\sin(\eta\tau + \psi(\tau)), \tag{4.30}$$

$$\dot{\overline{h}}(\tau) = a(\tau)\eta\cos(\eta\tau + \psi(\tau)) \tag{4.31}$$

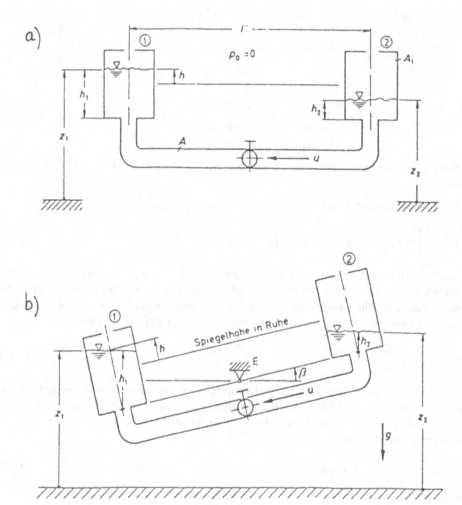

Figure 4.4: Schematic diagram of liquid damper
a) horizontal position,
b) inclined position

in the usual way leads to the approximate equations

$$\dot{a}(\tau) = -\frac{4}{3\pi}B\pi^2 a^2 - \frac{1}{2}E\sin\psi, \tag{4.32}$$

$$\dot{\psi}(\tau) = -\frac{\eta^2 - 1}{2\eta} - \frac{E}{2\eta\alpha}\cos\psi. \tag{4.33}$$

The stationary solution $a(\tau) \equiv a_s, \psi(t) \equiv \psi_s$ can easily be obtained from these equations as

$$a_s = \frac{3\pi}{8\sqrt{2}}\frac{1}{B\eta^2}\left\{-(1-\eta^2)^2 + \left[(1-\eta^2)^4 + \frac{256}{9\pi^2}E^2 B^2\eta^4\right]^{\frac{1}{2}}\right\}^{\frac{1}{2}}, \tag{4.34}$$

$$\sin\psi_s = -\frac{8}{3\eta}\frac{B}{E}a^2\eta^2, \tag{4.35}$$

$$\cos\psi_s = (1-\eta^2)\frac{a}{E}. \tag{4.36}$$

With the approximate analytic solution determined in this manner, the energy dissipated in the liquid damper can be computed, and the results of this simple calculation were compared to values measured in the laboratory. Fig. 4.5 shows a schematic representation of the laboratory setup in the Institut für Mechanik of the Darmstadt University.

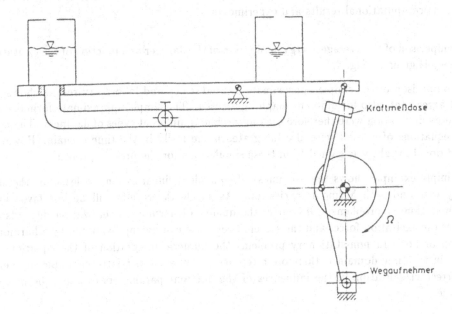

Figure 4.5: Schematic representation of experimental setup

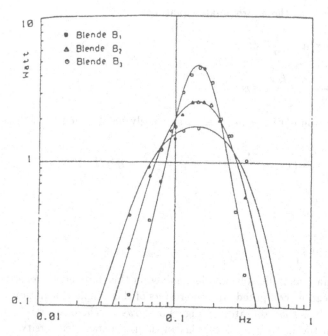

Figure 4.6: Average power dissipated in the liquid damper as a function of the excitation frequency; computational results and experiments

The comparison of the average energy dissipation of the damper as a function of the excitation frequency is given in Fig. 4.6.

This comparison of the simple approximate computations and laboratory results shows excellent agreement, also for different diaphragms used. The simple approximate formulae for the energy dissipation were therefore used for designing different types of dampers. The nonlinear equations of motion were also integrated numerically in the time domain. However, no additional insight was gained from these simulations for the present problem.

This simple example shows the usefulness of equivalent linearization, at least for the stationary case and with harmonic excitation. As a rule the results will be less favourable for non-stationary problems. Of course, the quality of harmonic linearization depends on whether the excitation forces and the system's response can be approximated by a harmonic function or not. In non-stationary problems the numeric integration of the equations of motion in the time domain is therefore often necessary; even in relatively simple problems the correct visualization of the influences of the different parameters can then be quite a challenge.

4.4 Final Remarks

As could be shown with the example from the previous section, simple approximate analytical solutions can be extremely useful in characterizing the general behaviour of nonlinear systems and showing the influence of the different design parameters. In this sense, approximate analytical solutions can be more useful than numerical solutions obtained from the numerical integration of the differential equations. Similar techniques are also used in the case of nonlinear systems subjected to stochastic loadings.

References

[1] **Hagedorn P.**: Nichtlinearitäten in der Schwingungstechnik. Einige Probleme und Lösungswege, Dynamische Probleme, Modellierung und Wirklichkeit, Mitteilung des Curt-Risch-Instituts der Universität Hannover, 1987, S. 1-22

[2] **Hagedorn P., Wallaschek J.**: On Equivalent Harmonic and Stochastic Linearization for Nonlinear Mechanical Elements, Proc. IUTAM Symp. on Nonlinear Stochastic Dynamic Engineering Systems, June 21-26, 1987, Innsbruck

[3] **Hagedorn P.**: Nonlinear Oscillations, Oxford University Press, Oxford 1988

5 Wind-excited Vibration of Transmission Lines

5.1 Introduction

In overhead transmission lines, different types of mechanical vibrations may occur. The most common type corresponds to wind-excited vibrations in the frequency range of 10 Hz to 50 Hz, caused by vortex shedding. Since these vibrations occur quite often, they may give rise to material fatigue, thus limiting the lifetime of the conductor. As capital investment is very high - even in small countries such as Germany the investment in high-voltage level transmission is of the same order of that directly connected with power generation - the problem of conductor vibration and fatigue merit being given the proper attention.

Since approximately 1930 the Stockbridge damper and similar devices have been used successfully to damp out these conductor vibrations. However, cases are known in which serious damage was caused to the conductor at the points of attachment of the dampers, i.e. at their clamps. Obviously this was due to the fact that the dynamic characteristics of damper and conductor had been improperly matched. Sufficient attention should be given to good modelling of this aspect of the problem. As we shall see, this can be done in different ways and at different levels. Only single conductors will be modelled in this paper, the problem of bundled conductors, which are widely used, will be treated in the next lecture.

Another very important type of vibrations in overhead transmission lines is the conductor "galloping". It is a low frequency ($f < 1$ Hz) vibrations which occurs only under very special meteorological conditions. It will be addressed shortly in the last part of the paper.

5.2   Vortex-excited Vibrations

5.2.1   The Mathematical Model of the Conductor

Figure 5.1 shows a typical span of a transmission line equipped with a damper which is usually mounted near the suspension clamp; there also may be more than one damper per conductor in each span. The span l, i.e. the distance between suspension towers, is usually in the order of 300 m to 1000 m. The conductor's sag is small, in general a few percent of the span. On the other hand, the frequencies under consideration (10 - 50 Hz) correspond to wave lengths of a few meters only, so that for the purpose of their study the sag can be disregarded, the conductor being modelled as a straight flexible continuous system. From observations it is known that although the conductor vibrations are not strictly planar, they occur predominantly in the vertical direction (see [2]). For this reason the dynamic behavior of overhead transmission line conductors can be conveniently studied in the plane.

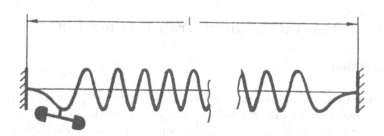

Figure 5.1: Overhead transmission line with damper

The mathematical model which is most often used for the conductor is the beam with bending stiffness EI under a large nominal force, i.e. a large tension T, whose transverse vibrations are described by

$$EIw^{IV}(x,t) - Tw''(x,t) + m\ddot{w}(x,t) = q(x,t) + d(w,\dot{w},t). \qquad (5.1)$$

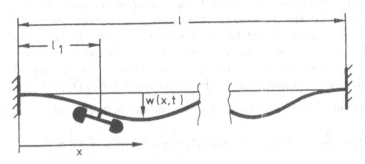

Figure 5.2: Conductor with damper

In (5.1), $w(x,t)$ is the transverse displacement at the location x of the conductor and at time t (see Fig. 5.2), m is the mass per unit length, $q(x,t)$ are the forces acting on the conductor due to wind action and vortex shedding and $d(w, \dot{w}, t)$ stands for the conductor's self-damping. The primes indicate differentiation with respect to x, while the dots stand for differentiation with respect to t. Equation (5.1) is valid for $x \neq l_1$; at the point $x = l_1$ the damper force has to be taken into account. The equation is usually solved for the boundary conditions

$$w(0,t) = w(l,t) = 0, \qquad w'(0,t) = w'(l,t) = 0 \tag{5.2}$$

which means that the suspension clamps are assumed as fixed during the vibrations. This is not necessarily so during actual vibrations, but it certainly is a case which may occur in reality (due to a symmetric span for example), and it is therefore taken as a reference in the calculations.

Supposing $q(x,t)$ and $d(w, \dot{w}, t)$ as known, as well as the damper force acting on the conductor, one would have to solve the boundary value problem formed by (5.1) and (5.2). The bending stiffness in (5.1) is a parameter which can usually only be determined experimentally due to the complicated structure of the conductors. They are formed by stranded wires which are neither completely free to slide with respect to the other wires nor do they form a rigid cross section. The bending stiffness consequently lies somewhere in between the corresponding two extreme values; it is small but essential if one wishes to calculate bending strains, which are responsible for the fatigue damages. It has, however, only negligible influence on the eigenfrequencies and eigenmodes of the free conductor vibrations (i.e. with $q(x,t) = 0, d(w, \dot{w}, t) = 0$) which are almost exactly those of a taut string (without bending stiffness) fixed at both ends. The eigenfrequencies are all the integer multiples of the first eigenfrequency, which in the case of a transmission line is typcially of the order of 0.1 Hz. This means that the frequency range of 10 Hz to 50 Hz corresponds to the interval between the 100th and the 500th eigenfrequencies of the conductor. All these modes then have to be modelled properly in the numerical solution of the nonlinear boundary value problems formed by (5.1) and (5.2), independently of the method of solution.

This implies a large number of elements if FEM techniques are used, or a high number of functions in the Ritz method for example. For bundled conductors, the numerical effort of course increases much more, almost becoming prohibitive, particularly in view of the fact that the wind forces are only modelled in a very rough manner, as we shall see later. There is therefore a strong interest in simplifying the model (without loosing essential information if possible). This goal is reached by using the fact that $\varepsilon = \sqrt{EI/Tl^2} \ll 1$.

With the dimensionless factor ε introduced above one can write (5.1) as

$$\varepsilon^2 w^{IV}(x,t) + \frac{1}{Tl^2}\left[-Tw''(x,t) + m\ddot{w}(x,t) - q(x,t) - d(w,\ddot{w},t)\right] = 0. \tag{5.3}$$

Recalling that ε^2 is a very small number, one may try to solve (5.3) by using perturbational calculus. This is in fact possible, it simplifies the problem enormously and gives excellent results. For $\varepsilon = 0$ equation (5.3) describes the vibrations of a flexible string (with zero bending stiffness); of course, the order of the partial differential equation is then lower, so that not all the four boundary conditions (5.2) can be satisfied. Only the first two boundary conditions are fulfilled by the zero-order solution to the problem.

In this zero-order problem, no bending strains can of course be calculated since there is no bending stiffness. With the zero-order solution known, higher-order approximations can be, however, constructed by standard perturbational techniques. Since the order of the partial differential equation (5.3) changes with ε^2 equal to or different from zero, this is a case of singular perturbations, and large bending strains only exist at the points at which the zero-order solution $w_0(x,t)$ has discontinuous slope, as is the case precisely at the points where concentrated forces act on the string, i.e. at the ends and at the point of attachment of the damper. At these places boundary layers occur, in which the bending strains decrease exponentially from a large value to almost zero. The locations of the bending boundary layers are shown in Fig. 5.3. It is at these points that fatigue may occur and where the bending strains should be checked.

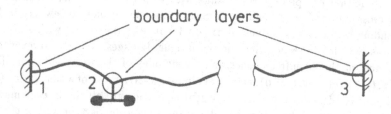

Figure 5.3: Bending boundary layers in a conductor

In [4] it is shown, for example, that the bending strain at the right-hand end of the conductor, i.e. in the boundary layer 3 is given by

$$w''(l,t) = \frac{T}{EI} w_0'(l^-,t), \qquad (5.4)$$

where w_0' is the slope at $x = l^-$ calculated from (5.3) with $\varepsilon^2 = 0$. Since $\sqrt{T/EI}$ is small compared to the length l and also to the wavelengths - it typically is of the order of a few centimeters while the wavelength is of the order of a few meters - the perturbational formulae like (5.4) give excellent results. The bending strains in the free field, i.e. far away from the bending boundary layers may be as small as one fiftieth of the bending strains in the boundary layers.

A comparison of the bending strains calculated with the full equations with those computed with the perturbational approach for realistic conductor parameters reveals that the difference is extremely small, usually below one percent. Due to this fact the perturbational formulae are now widely used since they considerably simplify the mathematical treatment.

5.2.2   Energy Balance

From many observations of vibrations caused by vortex shedding under steady wind conditions it is known that one frequency usually predominates at each wind speed. As will be seen later, this frequency is proportional to the speed of the wind blowing transversally to the conductor, which may vary slowly in time over a wide range. For given wind conditions the conductor can therefore be considered to oscillate with a single frequency. The eigenfrequencies of the conductor on the other hand are extremely dense, being spaced for example at 0.1 Hz intervals. Wind speeds leading to resonant conductor vibrations are therefore very close to speeds which correspond to anti-resonances. Also, the conductor length between two suspension clamps is usually not well known, and it is not even constant since it changes with temperature; the tension T in the conductor is subject to even more severe changes with temperature.

If a conductor oscillates, with a frequency of say, 32.1 Hz, it is for practical purposes not possible to distinguish whether it vibrates resonantly in the 322th, the 321th, or the 320th mode. However, this is not relevant since the wave-lengths corresponding to these neighboring modes are almost equal and so will be the maximum bending strains for given vibration amplitudes in the free field. Keeping in mind that a large part of the computational effort stems from the fact that a boundary value problem is being solved, one is therefore tempted to eliminate all the boundary conditions, at least during part of the calculations.

In fact, harmonic or quasi-harmonic mono-frequent vibrations being assumed, the vibration amplitudes completely determine the power introduced by the wind forces into the conductor. It turns out that the average energy level (averaged over one wave-length) is almost constant along most of the conductor. If this average energy or amplitude level is known for the free field, the bending strains for example at $x = l$ can easily be calculated. It is more or less obvious that the boundary conditions at $x = 0$ do not affect the bending strains at $x = l$ for the high modes being considered. For the infinite string the relation between wave-length λ and circular frequency ω is

$$\lambda = \omega \sqrt{T/m} \tag{5.5}$$

and the only effect of the conductor length and boundary layers is that the resonant frequencies are slightly shifted.

A simplified approach for the computation of the bending strains for a given wind speed is therefore obtained if one assumes that *any* frequency can be a resonant frequency for the conductor, computing the vibration amplitudes A in the free field from an energy balance of the type

$$P_W(A) = P_D(A) + P_C(A). \tag{5.6}$$

In (5.6) $P_W(A)$ is the power introduced into the conductor by the wind forces (for a given wind speed and frequency), $P_D(A)$ the power dissipated in the Stockbridge damper and $P_C(A)$ the power dissipated due to the conductor's self-damping properties. Expressions for $P_W(A)$ and $P_D(A)$ will be given in the following sections. The power $P_C(A)$ is small compared to $P_D(A)$ and will be disregarded in this lecture. Once the amplitude A has been determined from (5.6), the bending strains in the boundary layers can be computed by using simple perturbational formulae (see [4]).

An example of the results obtained with this simplified model is given in Fig. 5.4. It is seen that the vibration amplitude found in this manner as a function of the frequency (or wind speed) is the envelope to the amplitude curve from the boundary value problem. The resonant peaks are close to this envelope. In comparing both results one should keep in mind that equation (5.6) can be easily solved by using a pocket calculator, while the solution to the boundary value problem is a much more formidable task, further complicated by the fact that the wind forces cannot be precisely delineated.

5.2.3   Modelling of the Wind Forces

It is well-known that mainly vortex shedding is involved in the mechanism of excitation of the conductor vibrations in the frequency range under consideration. Surprisingly enough, relatively few data are available on the vibrations of a string or conductor even under the action of a laminar, steady transverse wind, particularly if the vibration amplitudes are as large as one conductor diameter, which is the case in transmission line vibrations.

Kármán vortex street model

If a stationary circular cylinder with diameter D is immersed in a planar uniform and stationary flow with velocity v, it is known that for a large range of Reynolds numbers vortices

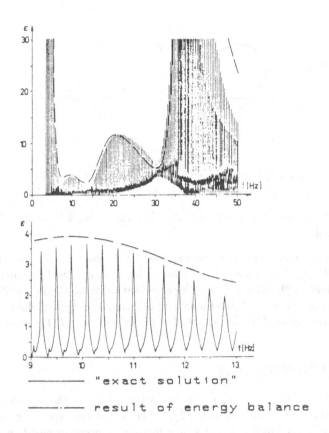

Figure 5.4: Bending strains: results of energy balance vs. boundary value problem

form alternatively at the upper and lower edge of the cylinder (see Fig. 5.5). The force $F(t)$ which the flowing medium applies to the cylinder, is periodic, and the term with the fundamental frequency very clearly dominates all the other terms of the FOURIER series of $F(t)$. This dominant part of $F(t)$ can be written as

$$F(t) = \frac{1}{2}\rho D L v^2 c_L \sin 2\pi f_s t \tag{5.7}$$

with ρ as the density of the flowing medium, D the cylinder's diameter, L its length, v the velocity of the unperturbed flow, c_L a coefficient which is of the order of 0.2 - 1.0 and f_s the Strouhal frequency given by

$$f_s = c_s v/D, \tag{5.8}$$

($c_s = 0.19$). The expression (5.7) is sometimes used in the computation of the wind-excited vibrations of overhead transmission lines. However, one should keep in mind that it is valid only for a stationary cylinder, or possibly also for a cylinder with small vibration amplitudes.

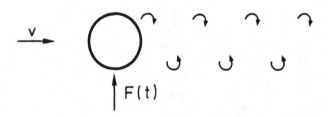

Figure 5.5: Vortex street forming at a circular cylinder

In the conductor vibrations, the amplitudes may become as large as D and (5.7) cannot be immediately applied.

Several researchers have measured the forces applied by the flow to a transversally oscillating cylinder, see [1], [3], [9]. The results differ considerably, and the most complete information seems to be contained in [9]. Also with a cylinder moved harmonically in transverse direction with amplitude A and frequency f, the fundamental harmonic in general dominates the other terms in the FOURIER series of the force applied to the cylinder and $F(t)$ can be approximated as

$$F(t) = \frac{1}{2}\rho D L v^2 c_L \sin(2\pi f t + \varphi),$$ (5.9)

where now c_L and also the phase angle φ between the force and the cylinder displacement depend on the ratios $a = A/D$ and $r = f/f_s$. The power corresponding to this force is positive for $0.9 < r < 1.3$ and $0 < a < 1$ approximately, it becomes negative otherwise. At $r = 1$ the "lock-in" phenomenon takes place, and force amplitudes may become much larger than in the Kármán vortex street observed at the stationary cylinder.

With the aid of (5.9) the power introduced by the wind in a transversally oscillating conductor can be estimated as

$$P_W = l f_s^3 D^4 \sum_{i=1}^{10} a_i \left(\frac{A}{D}\right)^i,$$ (5.10)

where a_i are coefficients which are determined with basis on the experiments. In deriving (5.10) it is taken into account that the amplitude is a function of x in the conductor's standing wave oscillation and in the expression of the power of the wind forces acting on the cylinder the cylinder's amplitude is substituted by $A \sin \pi x/\lambda$, followed by averaging over the wave-length.

5.2.4 Modelling of the Stockbridge Damper

Dampers of the Stockbridge and of similar types are formed by two rigid masses fixed at the ends of a "messenger cable" which is clamped to the conductor (see Fig. 5.6). The messenger cable consists of several steel wires and is built in such a way that a high amount of energy is dissipated in the deformation of these cables, i.e. when the end masses move. The dampers are usually placed near the suspension clamps, at a distance which should be less than one half wave-length of the highest-frequency-mode to be damped, so that they are never placed at a vibration node of the undamped cable.

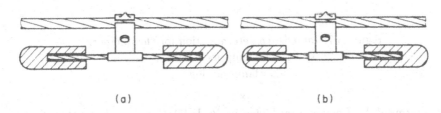

(a)   (b)

Figure 5.6: Typical dampers

Representation of the damper by its impedance matrix

In an actual line, the damper clamp usually executes a transverse translational as well as a rotational motion when in operation. In both types of motion, mechanical energy is dissipated in the damper. With the assumption of linearity the dynamic behavior of the damper can be characterized by the damper's driving point impedance matrix which can be measured in the laboratory. To this end two tests are carried out [5]: In the first one the damper clamp executes a translational harmonic oscillation with complex velocity amplitude $\hat{\dot{y}}$, in the second one, a rotational oscillation with angular velocity amplitude $\hat{\dot{\Theta}}$ and in both cases the moment and force amplitudes at the clamp $\hat{M}$ and $\hat{F}$ are measured. The elements Z_{ij} of the complex impedance matrix are then defined by

$$Z_{11} = \hat{F}_1/\hat{\dot{y}} \quad , \quad Z_{21} = \hat{M}_1/\hat{\dot{y}},$$
$$Z_{12} = \hat{F}_2/\hat{\dot{\Theta}} \quad , \quad Z_{22} = \hat{M}_2/\hat{\dot{\Theta}}, \tag{5.11}$$

where the index in the force and moment amplitudes characterizes the first and second experiment, respectively. Fig. 5.7 shows the impedance determination of the damper and the experimental set-up. It should be observed that in the experiments it may be convenient to use a pseudo-random excitation rather than a sweeping sine, since this is usually much faster.

The complete information with regard to the action of the damper can easily be deduced from the 2×2 impedance matrix $\mathbf{Z}(\omega)$, as shown in [5]. For a completely symmetric damper, the

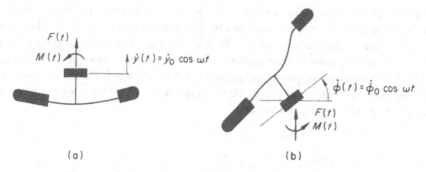

Figure 5.7: Impedance determination for the damper:
a) vertical clamp motion,
b) rotary clamp motion

off-diagonal terms of the $\mathbf{Z}$ matrix vanish identically. In [5] it was also shown that the moment impedance Z_{22} does not seriously affect the overall energy balance. For the calculation of the vibration amplitudes in the free field and the bending strains at the suspension clamp, it is therefore sufficient to consider only the term Z_{11} of the impedance matrix. The term Z_{22} is, however, important in the computation of the bending strains at the damper clamp.

The approach usually taken in studying the effect of the dampers on the conductor's vibrations is to consider only the single scalar impedance $Z_{11}(\omega)$, which suffices to correctly estimate the vibration amplitudes. An additional advantage of this approach is that the conductor's vibration analysis can be carried out with the string model (with zero bending stiffness), and the bending strains are computed a posteriori via singular perturbations. If the complete impedance matrix is used, concentrated moments are applied to the conductor, so that the beam has to be used to model the conductor.

The force acting on the damper located at $x = l_1$ can be written as (see Fig. 5.8)

$$F_D(t) = T \left[w'(l_1^+, t) - w'(l_1^-, t) \right].$$
(5.12)

If the conductor oscillates harmonically with frequency ω one has

$$w(x, t) = Re \left[W(x)e^{j\omega t} \right]$$
(5.13)

and

$$F_D(t) = \left[W'(l_1^+, t) - W'(l_1^-, t) \right] \cos \omega t.$$
(5.14)

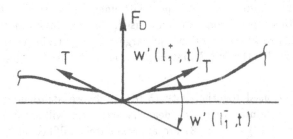

Figure 5.8: Force acting on the damper

On the other hand, the damper force can be related to the velocity $\dot{w}(l_1, t)$ by means of the damper impedance:

$$F_D(t) = Re\left[Z_{11}W(l_1)j\omega e^{j\omega t}\right] \tag{5.15}$$

so that $F_D(t)$ can be eliminated from (5.17) and (5.18). In [4] this approach is used to calculate the power P_D by the damper under given conditions.

In particular, one can compute which fraction of the vibration energy contained in a harmonic wave arriving from the free field towards the suspension clamp with the damper is reflected towards the free field and which part is absorbed by the damper (see Fig. 5.2). It can be expressed by means of the coefficient of absorption calculated in [4] as a function of Z_{11} and the cable parameters T, m, as well as the distance l_1 and the frequency ω. It turns out that the coefficient of absorption is equal to one, i.e. all the incoming vibration energy is absorbed for a certain damper impedance $(Z_{11})_{OPT}$ which can easily be calculated. If the optimal complex damper impedance is written as

$$(Z_{11})_{OPT} = |(Z_{11})_{OPT}|\, e^{j\beta} \tag{5.16}$$

one has

$$|(Z_{11})_{OPT}| = \frac{\sqrt{Tm}}{\sin\left(\omega \dfrac{l_1}{\sqrt{T/m}}\right)} \tag{5.17}$$

and

$$\beta_{OPT} = \frac{\omega l_1}{\sqrt{T/m}} - \frac{\pi}{2}. \tag{5.18}$$

It should be noted that the optimal damper impedance depends on the frequency and on the damper location. It is not advantageous to have a damper impedance constant over a large frequency range, as some manufacturers seem to believe. This goal is often aimed for by building asymmetric dampers, i.e. dampers with unequal "masses" at the two ends of the messenger cable. If this type of construction is used, the translation motion of the clamp is strongly coupled to its rotational motion, and large bending strains in the cable may occur at the damper clamp; also the full $\mathbf{Z}$ matrix has then to be used to describe the damper's dynamic behavior.

Relation (5.17) can be used together with (5.18) to choose a proper damper for a given conductor. Of course, a damper with such ideal characteristics will not be generally available, but if the conditions (5.17) and (5.18) are approximately fulfilled at least in part of the frequency range, the damper will work properly and not cause additional problems, as is sometimes the case if it is not well matched to the cable. From (5.17) it can also immediately be seen that the magnitude of the damper impedance should always be larger than the conductor impedance $\sqrt{Tm}$ for transverse waves.

5.3   Galloping, a Self-excited Vibration

While the vortex-excited vibrations discussed in the previous chapter are almost always present in a transmission line, usually with amplitudes well below one conductor diameter, galloping vibrations are only observed sporadically. As a rule they only occur in winter, when ice deposits on the conductor make it asymmetric and cause an aero-elastic instability. The general mechanism of this instability can be understood with a simple one-degree-of-freedom model.

Consider a rigid body in planar motion subjected to a stationary plane flow with velocity v_∞ (v_∞ is the velocity of the undisturbed flow at a distance far away from the body). The flowing medium then exerts forces and moments on the body, depending on the angle α (see Fig. 5.9).

The forces usually are resolved into two components: the *drag* D in the direction of v_∞ and the *lift* L orthogonal to the v_∞-direction (the influence of the resultant moment will not be considered here). The components L and D are determined experimentally in a wind tunnel and are represented by the drag coefficient c_D and the lift coefficient c_L, both of

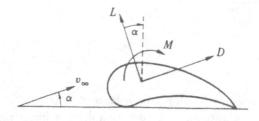

Figure 5.9: Flow-immersed rigid body with lift L, drag D, and angle of attack α.

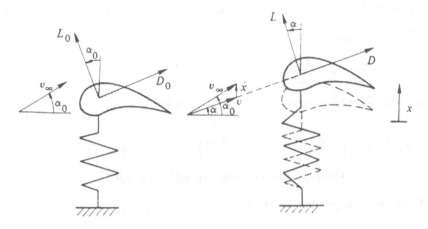

Figure 5.10: Flutter oscillations of an unstable profile

which depend on α:

$$L = \frac{1}{2}\rho c_L(\alpha)ldv_\infty^2, \qquad D = \frac{1}{2}\rho c_D(\alpha)ldv_\infty^2 \qquad (5.19)$$

where ρ is the mass density of the flowing medium, d is some characteristic diameter (which must be determined simultaneously with c_L and c_D), and l is the length of the profile under consideration. It is now assumed that only a vertical translation of the profile is admissible and that this motion is constrained by a linearly elastic spring with spring constant c (see Fig. 5.10).

If the vertical velocity of the profile is given by $\dot{x}$, then one has

$$v = \sqrt{\{(v_\infty \cos\alpha_0)^2 + (v_\infty \sin\alpha_0 - \dot{x})^2\}}. \qquad (5.20)$$

However, not only the magnitude of the velocity is changing but also the relative angle of attack which determines L and D; one has

$$\alpha = \arctan\left(\frac{v_\infty \sin\alpha_0 - \dot{x}}{v_\infty \cos\alpha_0}\right) = \arctan\left(\tan\alpha_0 - \frac{\dot{x}}{v_\infty \cos\alpha_0}\right). \qquad (5.21)$$

The lift and the drag again are given by (5.19) but with v_∞ replaced by v and with α satisfying the relationship (5.21). The equation of motion is

$$m\ddot{x} - L\cos\alpha - D\sin\alpha + cx = 0. \qquad (5.22)$$

For small oscillations about the equilibrium position with $x = x_0 + \overline{x}, \dot{x} = \dot{\overline{x}}$, one has

$$v^2 \approx v_\infty^2 - 2v_\infty \dot{\overline{x}} \sin \alpha_0, \tag{5.23}$$

$$\alpha = \alpha_0 + \overline{\alpha} \approx \alpha_0 - \frac{\dot{\overline{x}}}{v_\infty} \cos \alpha_0. \tag{5.24}$$

The linearized form of equation (5.22) thus is given by

$$m\ddot{\overline{x}} + \frac{\rho l d}{2} v_\infty \left[\left(\frac{d c_L}{d\alpha} + c_D \right) - \left(c_L - \frac{d c_D}{d\alpha} \right) \sin \alpha_0 \cos \alpha_0 + \right.$$
$$\left. + 2(c_L \cos \alpha_0 + c_D \sin \alpha_0) \sin \alpha_0 \right] \dot{\overline{x}} + c\overline{x} = 0 \tag{5.25}$$

with the equilibrium position defined by

$$x_0 = \frac{\rho l d}{2c} v_\infty^2 (c_L \cos \alpha_0 + c_D \sin \alpha_0). \tag{5.26}$$

Here, c_L and c_D and their derivatives in (5.21) and (5.22) must be evaluated at the angle of attack $\alpha = \alpha_0$. The linearized equation of motion thus becomes

$$m\ddot{\overline{x}} + \frac{\rho}{2} l d v_\infty b \dot{\overline{x}} + c\overline{x} = 0, \tag{5.27}$$

where b has been used to denote the bracket in equation (5.25). For the stability of the equilibrum position, one must have $b > 0$; for $b < 0$ the system is being excited. This stability criterion is simplified for $\alpha_0 = 0$ since one then has $b = d c_L/d\alpha + c_D$. In this form the stability criterion $d c_L/d\alpha + c_D > 0$ was already given by Den Hartog in 1932.

In winter asymmetric ice deposits may cause self-excited oscillations of the type described above. The mechanism can be much more complicated involving not only one-dimensional motion of the conductor's cross section, but also its rotation. In bundled conductors, which are quite common, the mechanical-mathematical model is still more complicated.

Galloping vibrations typically occur in one of the first eigenfrequencies of the conductor, that is with a frequency below 1 Hz but with very large amplitudes. Frequently, the amplitudes are of the order of several meters and lead to short circuits. Due to these short circuits transmission lines sometimes have to be taken out of operation, and this may have catastrophic consequences.

In contrast to the vortex-excited vibrations, which usually can be well controlled via dampers of the Stockbridge or of a different type, there is so far no universally accepted solution for controlling the galloping vibrations. The reason for this is not a lack of understanding of the complex mechanisms leading to this aero-dynamic instability but rather the difficulty to absorb vibration energy at a very low frequency under the technical constraints (small geometrical dimensions, low weight) given in an actual transmission line. The liquid-filled damper described in one of the previous lectures was one of many attempts to control these vibrations.

5.4 Conclusions

Wind-excited vibrations in overhead transmission lines involve very high costs associated both with conductor fatigue and with the catastrophic shut-downs due to short circuits. While the vortex-excited vibrations are usually well controlled via dampers, the same does not hold for galloping vibrations, although many passive and some active devices have been considered and tested.

References

[1] **Diana G., Falco M.**: On the Forces Transmitted to a Vibration Cylinder by a Blowing Fluid, Meccanica 6, 1971, p9-22

[2] **Doocy E.S., Hard A.R., Rawlins C.B., Ikegami R.**: Transmission Line Reference Book, Wind Induced Conductor Motion. Electric Power Research Institute, Palo Alto, Cal., 1979

[3] **Farquharson F.B., McHugh R.E.**: Wind Tunnel Investigation of conductor vibration with use of rigid models, AIEE Trans. PAS 75, p871-878, 1956

[4] **Hagedorn P.**: Ein einfaches Rechenmodell zur Berechnung winderregter Schwingungen an Hochspannungsleitungen mit Dämpfern, Ingenieur-Archiv, 49, 1980, p161-177

[5] **Hagedorn P.**: On the Computation of Damped Wind-Excited Vibrations of Overhead Transmission Lines, J. of Sound & Vibration, 83, 1982, p253-271

[6] **Hagedorn P., Kraus M.**: Aeolian Vibrations: Wind Energy Input Evaluated from Measurements on an Energized Transmission Line, IEEE Paper 90, SM 346-7 PWRD, presented at the IEEE Summer Power Meeting, Minneapolis, July 1990

[7] **Hagedorn P.**: Leiterseilschwingungen in Theorie und Praxis: Ein Überblick, etz-Report, 26, VDE-Verlag, Berlin 1990

[8] **Rawlins C.B.**: Power Imparted by Wind to a Model of a Vibrating Conductor, Electrical Products Division, ALCOA, Labs.; Massena, NY., 1982

[9] **Staubli T.**: Untersuchung der oszillierenden Kräfte am querangeströmten schwingenden Kreiszylinder, Doctoral Thesis, ETH Zürich, Switzerland, Diss. ETH 7322, 1983

6 The Control of Vibrations in Bundled Conductors via Self-Damping Spacers

6.1 Introduction

In overhead transmission lines the individual conductors of conductor bundles are kept apart by spacers for electrical reasons. The spacers are subject to different types of loads: small static forces due to different mechanical tensions in the conductors, extremely large loads for short time intervals (short circuits) and large quasi-periodic forces (galloping). In addition, they are regularly exposed to aeolian vibrations of the conductors in the frequency range between 10 and 60 Hz. High mechanical stresses may develop in the conductors, not only near the suspension clamps but also in the neighborhood of the spacer clamps.

In order to limit the bending stresses in the conductors and to absorb part of the conductors' aeolian vibration energy, viscoelastic parts were incorporated in the spacer. These new "spacer dampers" were first mentioned in [1].

Few papers have so far been devoted to the mathematical-mechanical modeling of spacer dampers in bundled conductors [2] - [5]. In the ideal case, the spacer dampers would damp the aeolian vibrations of bundled conductors to acceptable values in all subspans. Additional damping devices (e.g. Stockbridge dampers) would then be rendered superfluous. It is not yet clear under which conditions this goal can actually be achieved completely by the spacer dampers on the market. The practical evaluation of the performance of spacer dampers via field tests is being carried out by utilities and manufacturers. This is time consuming, and long term experiments have to be carried out under different meteorological and mechanical conditions.

It is of course clear, that a mathematical-mechanical model of a bundled line with spacer dampers can neither completely substitute the field tests nor the laboratory experiments. However, it can be a valuable tool in the design of spacer dampers and in deciding if for a given line sufficient damping can be achieved without additional damping devices. Such a model may also be useful in definig the minimum number of spacers dampers per span as well as their spacing.

In the present paper we develop a mathematical model for bundled conductors with spacer dampers with regard to aeolian vibrations only. For aeolian vibrations the individual conductors in the first approximation can be mechanically modelled as taut strings and described by the wave equation. The conductors' bending stiffness will not be taken into account for simplicity - it can however be included without problems. The conductor vibrations are assumed to be harmonic, and the spacer dampers are modeled as linear systems.

Mathematical models taking into account a complete span of a bundled conductor with spacer dampers, lead to a large system of (differential) equations amenable to numerical treatment only. Little insight will be gained by such numerical studies because of the system's

Figure 6.1: Spacer damper for a four bundle

sensitivity to small parameter variations. A different approach was therefore proposed in [6] for twin bundles and further developed in [7]. In this approach, the energy dissipated in a spacer damper of a bundle of four conductors is computed by modeling a single spacer with semi-infinite conductors attached at its clamps.

The design of spacer dampers may vary considerably from one manufacturer to the other. Most spacer dampers however have in common that their elasticity and damping depends on rubber elements. Such a spacer damper for a bundle of four conductors is depicted in Fig.6.1. In this particular design, the spacer damper consists of an aluminum frame, which can be considered rigid for the purpose of our considerations. Arms are attached to the central body by means of special flexible joints containing rubber elements. At the free end, each of these arms holds a clamp, to which the particular conductors are attached.

It is usually assumed that the aeolian vibrations of single conductors occur predominantly in a vertical plane, the wind blowing transversally in a horizontal direction giving rise to vertical pulsating forces on a conductor. In bundled conductors the situation is more complicated. Not only is the vortex shedding different from the single conductor case, because the vortices shedding from the windward conductor will hit the leeward conductor, but also the spacer damper itself complicates the situation further. Clearly, a design like the one shown in Fig.6.1 will provide coupling of the vibrations in the different directions. I.e. a vertical distributed and pulsating force according to Fig.6.2 will cause rotation of the arms, together with a compression of the flexible joints in the arm's longitudinal direction. The conductor will not move purely in a vertical plane.

In each of the four conductors the vibrations can be described in terms of travelling waves both in the horizontal and in the vertical plane. In both planes there are harmonic waves travelling from each side towards the clamp. These 16 incident harmonic waves may possibly arrive with different phase angles. Part of the vibration energy of the incident waves is absorbed in the spacer damper, the remaining part being reflected again in the form of

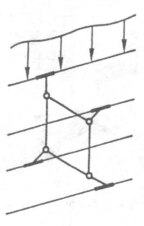

Figure 6.2: Bundle of four conductors

harmonic waves. This gives a total of 16 incident and 16 reflected waves. A possible criterion for the optimization of a spacer damper is the maximization of the energy dissipation, and this problem will be discussed in detail in what follows.

6.2   The Spacer Damper in a Bundle

Fig.6.3 represents the spacer damper of Fig.6.1 with the forces acting at the clamp. In this figure only one of the four arms is shown attached to the central frame via a viscoelastic joint at the point P. The forces in the directions transverse to the conductor are denoted by f_i, $i = 1, 2, 3, 4$, while the forces in the axial direction of the conductor are g_s, $s = 1, 2$. Since there are four arms, there is a total of 16 transverse force components $f_1, f_2, ..., f_{16}$ and 8 axial components $g_1, g_2, ..., g_8$. The conductor vibrations in the direction of $f_1, f_2, ...$ will be denoted by $w_1(x, t), w_2(x, t), ...$, respectively, and are described by the wave equation

$$\rho A_c \ddot{w}_i(x, t) = T_i w_i(x, t)'', \qquad i = 1, 2, ...16, \tag{6.1}$$

ρA_c being the conductor's mass per unit length and T_i its tension (normal force). Clearly $T_1 = T_2 = T_3 = T_4$ since the first four indices refer to the same conductor. The individual conductors may, however, be subject to different tensions so that possibly $T_4 \neq T_5$, etc. The harmonic travelling waves of the conductors are of the type

$$w_i(x, t) = Re\left[(W_i^+ e^{+jk_ix} + W_i^- e^{-jk_ix})e^{j\Omega t}\right], \qquad i = 1, 2, ..., 16, \tag{6.2}$$

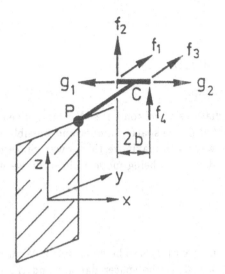

Figure 6.3: Spacer damper with forces acting at the clamp

where Ω is the angular frequency, k_i the wave number and $\Omega/k_i = \sqrt{T_i/\rho A_c}$ the wave velocity in the conductor.

The $W_i^{\pm}$ are the complex amplitudes of the waves travelling in the conductors in the positive and negative x-direction respectively. These complex amplitudes will be split up into the complex amplitudes of the incoming waves W_{Ii}, $i = 1, 2, \ldots, 16$, travelling towards the spacer clamp in a given section of the conductor, and the complex amplitudes W_{Li}, $i = 1, 2, \ldots, 16$, of the waves leaving the clamp. Refering to Fig.6.3 we have for example

$$
\begin{aligned}
W_{I1} &= W_1^+, & W_{I2} = W_2^+, \\
W_{L1} &= W_1^-, & W_{L2} = W_2^-,
\end{aligned}
\tag{6.3}
$$

since in the conductor section characterized by g_1 incoming waves travel in the direction of positive x, while the waves leaving the clamp travel in the direction of negative x. Similarly, in the conductor section characterized by g_2 we have the incoming waves travelling in the direction of *negative x*, so that

$$
\begin{aligned}
W_{I3} &= W_3^-, & W_{I4} = W_4^-, \\
W_{L3} &= W_3^+, & W_{L4} = W_4^+.
\end{aligned}
\tag{6.4}
$$

The vector of the complex amplitudes of the incoming waves is defined as

$$
\mathbf{W}_I = (W_{I1}, W_{I2}, \ldots, W_{I16})^T
\tag{6.5}
$$

and similarly we write

$$\mathbf{W}_L = (W_{L1}, W_{L2}, \ldots, W_{L16})^T \tag{6.6}$$

for the waves leaving the clamp.

The first goal of our calculations is to compute $\mathbf{W}_L$ for a given $\mathbf{W}_I$. We can than also calculate the power dissipated in the spacer damper and possibly optimize the spacer damper's parameters in such a way as to maximize this power. Since the system formed by the conductors and the spacer damper is being modeled linearly, we will have a relation of the type

$$\mathbf{W}_L = \mathbf{F}(\Omega)\mathbf{W}_I, \tag{6.7}$$

where the complex 16×16 matrix $\mathbf{F}(\Omega)$ is to be determined. With regard to the waves under consideration the system formed by the spacer damper and the conductors is therefore a "32 port" as shown in Fig.6.4.

It turns out that the matrix $\mathbf{F}(\Omega)$ can easily be calculated if the conductor parameters and the complex impedance matrix $\mathbf{Z}(\Omega)$ of the spacer damper are known. This matrix describes the relation between the complex amplitudes of the forces

$$\hat{\mathbf{f}} = (\hat{f}_1, \hat{f}_2, \ldots, \hat{f}_{16})^T \tag{6.8}$$

shown in Fig.6.3 and the complex amplitudes

$$\hat{\mathbf{w}} = (\hat{w}_1, \hat{w}_2, \ldots, \hat{w}_{16})^T \tag{6.9}$$

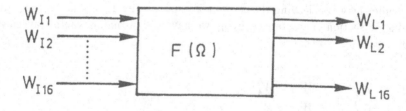

Figure 6.4: 32-port representing a four boundle spacer damper

of the velocities of the points of application of these forces in the respective directions during harmonic motion:

$$\hat{\mathbf{f}} = \mathbf{Z}(\Omega)\hat{\mathbf{w}}. \tag{6.10}$$

In what follows, we assume for the moment the complex 16×16 matrix is known; in a later section we will examine how it can be obtained.

Since we wish to use (6.10) to compute $\mathbf{W}_L$ for a given $\mathbf{W}_I$, we must first relate the wave motion in the conductor to the forces acting at the clamp. The force $f_1(t)$ in Fig.6.3 for example, is given by

$$f_1(t) = T_1 w_1'(-b, t), \tag{6.11}$$

$2b$ being the length of the clamp. With (6.2) this results in

$$f_1(t) = T_1 Re\left[jk_1(W_1^+ e^{-jk_1 b} - W_1^- e^{+jk_1 b})e^{j\Omega t}\right], \tag{6.12}$$

so that the complex force amplitude $\hat{f}_1$ is

$$\hat{f}_1 = T_1 j k_1 (W_1^+ e^{-jk_1 b} - W_1^- e^{jk_1 b}). \tag{6.13}$$

Similarly one has

$$\hat{f}_2 = T_2 j k_2 (W_2^+ e^{-jk_2 b} - W_2^- e^{jk_2 b}). \tag{6.14}$$

While the forces f_1, f_2 act at $x = -b$, the forces f_3, f_4 are applied at $x = +b$, so that

$$\hat{f}_3 = T_3 j k_3 (W_3^+ e^{jk_3 b} - W_3^- e^{-jk_3 b}), \tag{6.15}$$

$$\hat{f}_4 = T_4 j k_4 (W_4^+ e^{jk_4 b} - W_4^- e^{-jk_4 b}). \tag{6.16}$$

The complex amplitudes $\hat{f}_i$, $i = 5, 6, \ldots, 16$, are given by analogous expressions. These force amplitudes are therefore linear combinations of the amplitudes of the incoming waves and of the waves leaving the clamps, i.e.

$$\hat{\mathbf{f}} = \mathbf{A}\mathbf{W}_I + \mathbf{B}\mathbf{W}_L, \tag{6.17}$$

where the complex 16×16 matrices $\mathbf{A}, \mathbf{B}$ depending on the $k_i = \Omega/\sqrt{T_i/\rho A_c}$, $i = 1, 2, \ldots, 16$, are easily obtained from the equations similar to (6.12) - (6.16).

Next we turn to the velocity amplitudes appearing in (6.10). The velocity component $\dot{w}_1$ of the point of application of f_1 in the direction of this force is

$$\dot{w}_1(t) = \dot{w}(-b, t), \tag{6.18}$$

and with (6.2) this gives

$$\dot{w}_1(t) = Re\left[j\Omega(W_1^+ e^{-jk_1 b} + W_1^- e^{jk_1 b})e^{j\Omega t}\right], \tag{6.19}$$

$$\hat{\dot{w}}_1 = j\Omega(W_1^+ e^{-jk_1 b} + W_1^- e^{jk_1 b}). \tag{6.20}$$

Similarly,

$$\hat{\dot{w}}_2 = j\Omega(W_2^+ e^{-jk_2 b} + W_2^- e^{jk_2 b}), \tag{6.21}$$

$$\hat{\dot{w}}_3 = j\Omega(W_3^+ e^{jk_3 b} + W_3^- e^{-jk_3 b}), \tag{6.22}$$

$$\hat{\dot{w}}_4 = j\Omega(W_4^+ e^{jk_4 b} + W_4^- e^{-jk_4 b}), \tag{6.23}$$

etc. Also the vector $\hat{\dot{w}}$ can therefore be written as a linear combination of W_I and W_L :

$$\hat{\dot{w}} = EW_I + HW_L. \tag{6.24}$$

The complex 16×16 matrices D and E are again easily found from equations of the type (6.20) - (6.23).

Substituting (6.17) and (6.24) into (6.10) results in

$$AW_I + BW_L = Z(\Omega)(EW_I + HW_L), \tag{6.25}$$

which can be solved for W_L :

$$W_L = (B - HZ(\Omega))^{-1}(-A + EZ(\Omega))W_I. \tag{6.26}$$

In (6.26) not only the matrix Z but also A, B, E, H depend on Ω. The matrix $F(\Omega)$ in (6.7) is therefore given by

$$F = (B - HZ)^{-1}(-A + EZ). \tag{6.27}$$

We are now in a position to calculate the complex amplitudes of the waves leaving the spacer damper for a given $\mathbf{W}_I$. Next we wish to compute the power dissipated in the spacer damper for given $\mathbf{W}_I$.

The power dissipated in the spacer damper averaged with respect to time is

$$P_S = -\frac{1}{2} Re(\hat{\mathbf{f}}^* \hat{\mathbf{w}}), \tag{6.28}$$

with the asterix meaning "complex conjugate transposed". We first write $\hat{\mathbf{f}}$ and $\hat{\mathbf{w}}$ as linear functions of $\mathbf{W}_I$. With (6.24) and (6.26) we obtain

$$\hat{\mathbf{w}} = \left[\mathbf{E} + \mathbf{H}(\mathbf{B} - \mathbf{HZ})^{-1}(-\mathbf{A} + \mathbf{EZ})\right] \mathbf{W}_I, \tag{6.29}$$

which can also be written as

$$\hat{\mathbf{w}} = \mathbf{K}(\Omega)\mathbf{W}_I. \tag{6.30}$$

Using (6.10) and (6.30), (6.28) finally leads to

$$P_S = -\mathbf{W}_I^* \mathbf{G}(\Omega)\mathbf{W}_I, \tag{6.31}$$

where the 16 × 16 complex hermitean matrix $\mathbf{G}(\Omega)$ is

$$\mathbf{G}(\Omega) = \frac{1}{2}\mathbf{K}^*(\Omega)\mathbf{Z}^*(\Omega)\mathbf{K}(\Omega), \tag{6.32}$$

with

$$\mathbf{K} = \mathbf{E} + \mathbf{H}(\mathbf{B} - \mathbf{HZ})^{-1}(-\mathbf{A} + \mathbf{EZ}). \tag{6.33}$$

The problem of computing the power dissipated in the spacer damper for given incident waves has therefore been solved (under the assumption of a given impedance matrix for the spacer damper).

The power transported towards the spacer damper by the incident waves, i.e. the "incident power" is given by

$$P_I = \mathbf{W}_I^* \mathbf{T}(\Omega)\mathbf{W}_I, \tag{6.34}$$

where $\mathbf{T}$ is a 16 × 16 real diagonal matrix with diagonal elements

$$T_{ii} = \frac{1}{2}\Omega^2 \sqrt{\rho A_c T_i}, \qquad i = 1, 2, \dots 16. \tag{6.35}$$

We define an absorption coefficient q, $0 \leq q \leq 1$, as in [7], as the quotient of the dissipated by the incident power

$$q = \frac{P_S}{P_I} = -\frac{\mathbf{W}_I^* \mathbf{G}(\Omega)\mathbf{W}_I}{\mathbf{W}_I^* \mathbf{T}(\Omega)\mathbf{W}_I}. \tag{6.36}$$

This coeffient is a homogeneous function of degree zero in $\mathbf{W}_I$; it assumes different values for different combinations of the complex amplitudes $W_{I1}, W_{I2}, \dots, W_{I16}$ of the waves incident along the different conductor sections.

The phase angles between the different waves are not known a priori, the *minimum* value q_{min} of this quotient can be used as a measure for the efficiency of the spacer damper at a given value of the frequency Ω. It gives the fraction of incident power which is dissipated in the spacer damper for the worst case of the phase angles between the different incident waves. The design of the spacer damper should therefore be optimized in such a way as to maximize q_{min} for each Ω.

The minimum value of the absorption coefficient defined by (6.36) not only can be used in the optimization of the spacer damper design but also in the energy balance as outlined in [7]. Using energy balance techniques it is possible to estimate the number of spacer dampers necessary per span for a given limit of vibration angles or nominal bending strains at the clamp in the conductors. This question will not be further adressed in the present lecture.

The function $q(\mathbf{W}_I)$ defined by (6.36) assumes a somewhat simpler form if all the conductors are subjected to the same tension T. In this case the diagonal matrix $\mathbf{T}$ defined by (6.35) is proportional to the identity matrix

$$\mathbf{T} = \frac{1}{2}\Omega^2 \sqrt{\rho A_c T} \ \ \mathbf{I} \tag{6.37}$$

and the absorption coefficient is

$$q = -\frac{2}{\Omega^2 \sqrt{\rho A_c T}} \frac{\mathbf{W}_I^* \mathbf{G}(\Omega)\mathbf{W}_I}{\mathbf{W}_I^* \mathbf{W}_I}. \tag{6.38}$$

For a given value of Ω the minimum of q corresponds to the lowest eigenvalue $\alpha(\Omega)$ of the hermitean matrix $-\mathbf{G}(\Omega)$, i.e.

$$q_{min} = \frac{2\alpha(\Omega)}{\Omega^2 \sqrt{\rho A_c T}}. \tag{6.39}$$

The minimization of $\mathbf{q}(\mathbf{W}_I)$ in this case therefore reduces to the solution of a standard eigenvalue problem for a hermitean matrix.

6.3 Calculation of the Impedance Matrix

In the previous section the impedance matrix $\mathbf{Z}(\Omega)$ of the spacer damper was assumed to be known. In principle it can be found directly from laboratory experiments or calculated with the known design parameters of the spacer [7]. The direct experimental determination of the 16×16 impedance matrix is however a formidable task. Moreover, in the optimization of the design the dependence of $\mathbf{Z}(\Omega)$ on the design parameters is needed. In what follows we therefore give a description of the analytical determination of $\mathbf{Z}(\Omega)$.

The spacer damper in Fig.6.3 consists of a central rigid body connected to 4 arms by visco-elastic hinges. The spacer damper is therefore a system formed by 5 rigid bodies, i.e. it is a system of $5 \times 6 = 30$ degrees of freedom. The motion of each of the 5 bodies is ruled by the linear and angular momentum equations

$$m_i \ddot{\vec{r}}_i = \sum_j \vec{p}_{i,j} \tag{6.40}$$

$$\frac{d}{dt}(\theta_i \vec{\omega}_i) = \sum_j \vec{v}_{i,j}, \qquad i = 1,2,3,4,5. \tag{6.41}$$

Here m_i is the mass of the i-th body, $\ddot{\vec{r}}_i$ the acceleration of its center of mass with respect to an inertial frame of reference, θ_i its central inertia tensor and $\vec{\omega}_i$ its angular velocity with respect to an inertial frame. The $\vec{p}_{i,j}$ are the forces acting on the i-th body and the $\vec{v}_{i,j}$ the moments about the center of mass of all the forces on the i-th body; the sums on the right hand side of (6.40) extend over all these forces and moments, respectively.

The configuration of the system is represented at each instant by the values of the generalized coordinates

$$\mathbf{q} = (q_1, q_2, \ldots, q_{30})^T, \tag{6.42}$$

which can be arbitrarily chosen. Typically, for each body three linear coordinates and three angles will be used. Since we are only interested in oscillations of small amplitudes, the three rotation angles of each of the 5 bodies about the x-, y- and z-axes shown in Fig.6.3 can be used as generalized coordinates, and there is no need to use EULER angles or an equivalent representation. In the equations of motion linearized about the equilibrium position, the left hand sides of (6.40) will simply be linear combinations of the $\ddot{q}_i$, $i = 1,2,\ldots,30$. The forces at the viscoelastic joints in linearized form can be written as linear combinations of q_i and $\dot{q}_i$, $i = 1,2,\ldots,30$.

With the system in equilibrium, the transverse forces f_i, $i = 1, 2, \ldots, 16$ vanish, while the axial forces g_i, $i = 1, 2, \ldots, 8$ are equal to the respective tensions in the cables. For the oscillating system we therefore have

$$g_i = T_i + \overline{g}_i, \qquad i = 1, 2, \ldots, 8, \tag{6.43}$$

where the $\overline{g}_i$, $i = 1, 2, \ldots, 8$, are small of first order in the $\mathbf{q}$. If the arm of Fig.6.3 is rotated about his axis PC, it is clear that T_1 and T_2 will produce a restoring torque of first order in $\mathbf{q}$, while the torque produced by $\overline{g}_1$ and $\overline{g}_2$ will be of second order.

The linearized equations of motion of the spacer damper are therefore of the type

$$\mathbf{M\ddot{q}} + \mathbf{D\dot{q}} + \mathbf{Cq} = \mathbf{Lf} + \mathbf{Nu}, \tag{6.44}$$

with $\mathbf{M}$, $\mathbf{D}$, $\mathbf{C}$ real symmetric 30×30 matrices, $\mathbf{L}$ a real 16×30 matrix and $\mathbf{N}$ a real 4×30 matrix. As before,

$$\mathbf{f} = (f_1, f_2, \ldots, f_{16})^T \tag{6.45}$$

is the vector of the forces acting on the damper clamps in the directions transverse to the conductors of Fig.6.3 and

$$\mathbf{u} = (\overline{g}_1 - \overline{g}_2, \overline{g}_3 - \overline{g}_4, \overline{g}_5 - \overline{g}_6, \overline{g}_7 - \overline{g}_8)^T \tag{6.46}$$

contains the forces acting on the clamps in the conductors' axial direction. Only the differences of the $\overline{g}_i$, $i = 1, 2, \ldots, 8$ contained in (6.46) appear in the linearized equations of motion, if we assume that these terms are themselves small of the first order. In (6.44) the matrix $\mathbf{M}$ is easily obtained from the mass geometry of the spacer damper. The matrices $\mathbf{D}$ and $\mathbf{C}$, since they describe how the forces and moments in the viscoelastic joints depend on the displacements and velocities, contain information both on the viscoelastic properties of the joints and on the geometry of the spacer damper. In addition the matrix $\mathbf{C}$ contains the coefficients of the restoring torques due to T_i, $i = 1, 2, \ldots, 8$. The matrices $\mathbf{L}$ and $\mathbf{N}$ contain geometric information only.

The clamps are however not free to move in the axial direction of the conductors. We assume that the axial stiffness of the conductor is such that the clamps are rigidly fixed in this direction. This corresponds to 4 conditions of constraint, which in linearized form are written as

$$\mathbf{Aq} = 0, \tag{6.47}$$

where the 30×4 real matrix $\mathbf{A}$ contains geometric parameters only. The constraint equation (6.47) can be used to eliminate $\mathbf{u}$ from (6.44). Since we are interested only in harmonic oscillations of the type

$$\mathbf{q}(t) = Re\left(\hat{\mathbf{q}}e^{j\Omega t}\right) \tag{6.48}$$

with complex amplitude $\hat{\mathbf{q}}$, (6.44) can be reduced to

$$(-\Omega^2 \mathbf{M} + j\Omega \mathbf{D} + \mathbf{C})\hat{\mathbf{q}} = \mathbf{L}\hat{\mathbf{f}} + \mathbf{N}\hat{\mathbf{u}}. \tag{6.49}$$

In this equation $\hat{\mathbf{f}}$ and $\hat{\mathbf{u}}$ are the complex force amplitudes corresponding to (6.45), (6.46). Solving (6.49) for $\hat{\mathbf{q}}$ gives

$$\hat{\mathbf{q}} = (-\Omega^2 \mathbf{M} + j\Omega \mathbf{D} + \mathbf{C})^{-1}(\mathbf{L}\hat{\mathbf{f}} + \mathbf{N}\hat{\mathbf{u}}) \tag{6.50}$$

and substituting into

$$\mathbf{A}\hat{\mathbf{q}} = 0 \tag{6.51}$$

yields

$$\mathbf{A}(-\Omega^2 \mathbf{M} + j\Omega \mathbf{D} + \mathbf{C})^{-1}(\mathbf{L}\hat{\mathbf{f}} + \mathbf{N}\hat{\mathbf{u}}) = 0. \tag{6.52}$$

These are 4 scalar equations which can be solved for $\hat{\mathbf{u}}$:

$$\hat{\mathbf{u}} = (-\mathbf{A}(-\Omega^2 \mathbf{M} + j\Omega \mathbf{D} + \mathbf{C})^{-1}\mathbf{N})^{-1}(\mathbf{A}(\Omega^2 \mathbf{M} + j\Omega \mathbf{D} + \mathbf{C})^{-1}\mathbf{L})\hat{\mathbf{f}}. \tag{6.53}$$

The vector $\hat{\mathbf{u}}$ is therefore known as a linear function of $\hat{\mathbf{f}}$, and if it is substituted in (6.49) this equation can be solved for $\hat{\mathbf{q}}$ giving

$$\hat{\mathbf{q}} = \mathbf{J}(\Omega)\hat{\mathbf{f}}. \tag{6.54}$$

The explicit form of the complex 30×16 matrix $\mathbf{J}(\Omega)$ is not given here for simplicity, but it follows easily from the equations above.

The geometric displacements w_i in the directions of the f_i, $i = 1, 2, \ldots, 16$ at the end points of the clamps can be written as linear combinations of the q_j, $j = 1, 2, \ldots, 30$ such as

$$\mathbf{w} = \mathbf{Q}\mathbf{q}, \tag{6.55}$$

with

$$\mathbf{w} = (w_1, w_2, \ldots, w_{16})^T, \tag{6.56}$$

the real 16×30 matrix $\mathbf{Q}$ again containing geometric information only. For harmonic oscillations (6.55) can be written as

$$\hat{\mathbf{w}} = \mathbf{Q}\hat{\mathbf{q}}, \tag{6.57}$$

which together with (6.54) leads to

$$\hat{\mathbf{w}} = \mathbf{Q}\mathbf{J}(\Omega)\hat{\mathbf{f}}, \tag{6.58}$$

or

$$\hat{\mathbf{f}} = (\mathbf{Q}\mathbf{J}(\Omega))^{-1}\hat{\mathbf{w}}. \tag{6.59}$$

This finally gives

$$\hat{\mathbf{f}} = \frac{1}{j\Omega}(\mathbf{Q}\mathbf{J}(\Omega))^{-1}\hat{\mathbf{w}}, \tag{6.60}$$

so that the complex impedance matrix sought is

$$\mathbf{Z}(\Omega) = \frac{1}{j\Omega}(\mathbf{Q}\mathbf{J}(\Omega))^{-1}. \tag{6.61}$$

Note that $\mathbf{Z}(\Omega)$ contains information not only on the spacer damper but also on the conductor tensions.

In Darmstadt a computer program was written for the computation of the impedance matrix of the spacer damper depicted in Fig.6.1. The expression for $\mathbf{Z}(\Omega)$ is generated in algebraic form using MATHEMATICA 2.0. This is convenient for parameter studies and for the optimization of the spacer damper, since the dependence on the design parameters is explicitly given. The mechanical stiffness and damping parameters of the viscoelastic joints were experimentally determined in the laboratory. Since the damping is provided by rubber elements, the damping matrix $\mathbf{D}$ in (6.44) is frequency dependent. The term "viscoelastic" so far used to describe the properties of the joint is therefore misleading. In the present context it simply means that the joint has linear elastic and damping properties.

For the spacer damper shown in Fig.6.3 with given conductors, the expressions for the impedance matrix $\mathbf{Z}(\Omega)$ and for the absorption coefficient q were computed as shown above. In the near future parameter studies will be carried out with the intent to optimize the design of the spacer damper.

6.4   Final Remarks

In this paper a mathematical model was developed for a conductor bundle with a single spacer damper. Transverse waves in the individual conductors are travelling towards the spacer damper and then are partially reflected. A coefficient of absorption is computed via the solution of the eigenvalue problem of a hermitean 16×16 matrix, which was obtained after lengthy but simple calculations. The dependence of this matrix on the spacer damper's design parameters can be given in explicit form. It is therefore more practical to optimize the spacer damper for a given conductor in this manner then by solving the eigenvalue problem of a complete span of bundled conductors with many spacer dampers. This latter model leads to a numerically poorly conditioned problem of a much larger order of magnitude.

The question of optimal location of the spacer dampers in a bundle of given length is of course an important problem, which is being studied separately, the main criterion for optimal spacer placement being the pervasiveness of damping.

Acknowledgement: The support of *Richard Bergner GmbH + Co., Schwabach, Germany*, is greatfully acknowledged.

References

[1] **Edwards A.T, Boyd J.M.**: Bundle-Conductor-Spacer Design Requirements and Development of "Spacer-Vibration Damper", IEEE Transactions on Power Apparatus and Systems, Paper 63-1075, 1965

[2] **Claren R., Diana G., Giordana F., Massa E.**: The vibration of transmission line conductor bundles, IEEE paper, 71 TP 158-PWR, 1971

[3] **Simpson A.**: Determination of the natural frequencies of multiconductor overhead transmission lines, Journal of Sound and Vibration, **20** (1972), pp. 417-449, 1972

[4] **Möcks L., Schmidt J.**: Bemessung und Anordnung von selbstdämpfenden Feldabstandhaltern in Bündelleitern, Elektrizitätswirtschaft, **87**, H.21, pp. 1044-1048, 1988

[5] **Bourdon P., Brunelle J., Lavigne, P.**: Methods of evaluating dynamic characteristics of dampers, spacers and spacer dampers, CAE Research Report, contract no. 145 T 327, 1987

[6] **Wallaschek J.**: Zur Dämpfung winderregter Schwingungen in den Bündelleitern elektrischer Freileitungen, ZAMM, **71**, pp. 300-303, 1991

[7] **Hagedorn P., Kraus M.**: On the performance of spacer dampers in bundled conductors, Paper submitted to the IEEE Summer Power Meeting, Seattle, July 1992

CHAPTER II

SEISMIC ISOLATION SYSTEMS:
INTRODUCTION AND OVERWIEW

M.C. Constantinou
State University of New York at Buffalo, Buffalo, NY, USA

ABSTRACT

Seismic isolation (also often referred to as base isolation) is a technique for mitigating the effects of earthquakes on structures through the introduction of flexibility and energy absorption capability. Various practical means for introducing this desired flexibility and energy absorption capability are described.

INTRODUCTION

Many methods have been proposed for achieving optimum performance of structures subjected to earthquake excitation. The conventional approach requires that structures passively resist earthquakes through a combination of strength, deformability, and energy absorption. The level of damping in these structures is typically very low and therefore the amount of energy dissipated during elastic behavior is very low. During strong earthquakes, these structures deform well beyond the elastic limit and remain intact only due to their ability to deform inelastically. The inelastic deformation takes the form of localized plastic hinges which results in increased flexibility and energy dissipation. Therefore, much of the earthquake energy is absorbed by the structure through localized damage of the lateral force resisting system. This is somewhat of a paradox in that the effects of earthquakes (i.e., structural damage) are counteracted by allowing structural damage.

An alternative approach to mitigating the hazardous effects of earthquakes begins with the consideration of the distribution of energy within a structure. During a seismic event, a finite quantity of energy is input into a structure. This input energy is transformed into both kinetic and potential (strain) energy which must be either absorbed or dissipated through heat. However, there is always some level of inherent damping which withdraws energy from the system and therefore reduces the amplitude of vibration until the motion ceases. The structural performance can be improved if a portion of the input energy can be absorbed, not by the structure itself, but by some type of supplemental "device". This is made clear by considering the conservation of energy relationship

$$E = E_k + E_s + E_h + E_d \tag{1}$$

where E is the absolute energy input from the earthquake motion, E_k is the absolute kinetic energy, E_s is the recoverable elastic strain energy, E_h is the irrecoverable energy dissipated by the structural system through inelastic or other forms of action, and E_d is the energy dissipated by supplemental damping devices. The absolute energy input, E, represents the work done by the total base shear force at the foundation on the ground (foundation) displacement. It, thus, contains the effect of the inertia forces of the structure.

In the conventional design approach, acceptable structural performance is accomplished by the occurrence of inelastic deformations. This has the direct effect of increasing energy E_h. It also has an indirect effect. The occurrence of inelastic deformations results in softening of the structural system which itself modifies the absolute input energy. In effect, the increased flexibility acts as a filter which reflects a portion of the earthquake energy.

The technique of seismic isolation (e.g., Buckle 1990, Kelly 1993, Mokha 1991, Constantinou 1991a) accomplishes the same task by the introduction, at the foundation of a structure, of a system which is characterized by flexibility and energy absorption capability. The flexibility alone, typically expressed by a period of the order of 2 seconds, is sufficient to reflect a major portion of the earthquake energy so that inelastic action does not occur. Energy dissipation in the isolation system is then useful in limiting the displacement response and in avoiding resonances. The reduction of bearing displacements in highly damped isolation systems is advantageous in bridges where small displacements result in short expansion joints and avoid the use of knock-off elements (Constantinou 1991a, 1992, 1993; Tsopelas 1994).

The reduction of bearing displacements in highly damped isolation systems result, typically, in reduction of the isolation system shear force. However, in hysteretic isolation systems (e.g. high damping rubber bearings, lead-rubber bearings and sliding isolation systems) high levels of energy dissipation lead to out-of-phase, high accelerations in the superstructure (that is, accelerations higher than those in a lightly damped isolated structure but still less than those in a non-isolated structure). This behavior is undesirable when the intent of seismic isolation is to protect expensive sensitive equipment in the isolated building (Kelly 1993). However when the intent of seismic isolation is to protect the structural system, high levels of hysteretic damping are beneficial.

Evidence for this behavior is provided in Figures 1 and 2. The presented results are from analyses of an 8-story isolated building supported by 45 isolators (Winters 1993). Each isolation bearing has bilinear hysteretic properties which characterize a wide range of elastomeric and sliding systems. Twelve isolation systems with effective isolation period in the range of 1.5 to 3s and effective damping ratio in the range of 0.06 to 0.37 of critical were analyzed. The analysis included the effects of bidirectional interaction and eccentricities. The input was represented by 9 pairs of actual earthquakes scaled to be representative of Seismic Zone 4 (California), soil type S_2 in accordance with the Uniform Building Code in the United States (ICBO 1991).

Figure 1 demonstrates the increase of bearing displacement with increasing period and the reduction of displacement with increasing damping. Figure 2 demonstrates the reduction of base shear force with increasing damping. Interestingly, the shear force in the upper stories is about the same for all levels of damping. However, the highly damped systems show a shear force distribution which is nearly constant with the height of the structure. This indicates higher mode, out-of-phase, response, which, typically, is accompanied by high floor accelerations. Nevertheless, moderately and highly damped isolation systems offer the advantage of lower bearing displacements, while structural shear forces are about the same or less than those of lightly damped systems.

The benefits of low bearing displacements, low structural forces and low floor accelerations may be realized with isolation systems which exhibit linear and viscous (not hysteretic behavior). This is demonstrated in the comparison of results in Figure 3. Such systems may be constructed by combining low damping elastomeric bearings with fluid viscous dampers (Constantinou 1992b). Such fluid dampers have been in use in the United States for the last 30 years in shock and vibration isolation applications, primarily in the military. Recently, such devices have been specified for use in the seismic isolation system of a hospital complex in California.

Modern seismic isolation systems incorporate energy dissipating mechanisms. Examples are high damping elastomeric bearings, lead plugs in elastomeric bearings, mild steel dampers, fluid viscous dampers, and friction in sliding bearings.

BASIC ELEMENTS OF SEISMIC ISOLATION SYSTEMS

Buckle, 1990 identified the basic elements in a practical isolation system:

(1) Flexibility to lengthen the period and produce the isolation effect,
(2) Energy dissipation capability to reduce displacements to practical design levels, and
(3) Means for providing rigidity under service loads such as wind.

MODERN SEISMIC ISOLATION SYSTEMS

Elastomeric bearings represent a common, but not the only, means for introducing flexibility into an isolated structure. Figure 4 shows the construction of one such bearing. It consists of thin layers of natural rubber which are vulcanized and bonded to steel plates. Rubber exhibits low shear modulus, typically $G = 0.5$ to 1 MPa at shear strain of about 50%, which is unaffected by the insertion of the steel plates. Effectively, the horizontal stiffness of the bearing is given by

$$K_H = \frac{AG}{\Sigma t} \qquad (2)$$

where A = bonded rubber area and Σt = total rubber thickness. However, the insertion of the steel plates greatly reduces the freedom of rubber to bulge so that the vertical stiffness of the bearing is large. The vertical stiffness is given by

$$K_v = \frac{AE_c}{\Sigma t} \qquad (3)$$

where E_c = compression modulus. Analytical-empirical relations for evaluating the compression modulus of elastomeric bearings have been included in specifications such as the AASHTO, 1991. An approximate expression for the compression modulus was derived by Chalhoub 1990 and Constantinou 1992c based on principles of mechanics and physically motivated assumptions:

$$E_c = \left(\frac{1}{6GS^2F} + \frac{4}{3K} \right)^{-1} \tag{4}$$

where S = shape factor, defined as the loaded rubber area divided by the rubber area which is free to bulge, K = bulk modulus, typically around 2000 MPa and F = factor which takes a value equal to unity for circular bearings and a value between 2/3 and 1 for circular bearings with a central hole (central holes are required in the manufacturing of large diameter bearings). Since the shape factor is in the range of 10 to almost 40, the vertical stiffness is between 400 and 1300 times greater than the horizontal stiffness. This is particularly important in preventing rocking motion of the isolated structure.

Elastomeric bearings made of low damping natural rubber (NRB) exhibit equivalent viscous damping of the order of 0.05 or less of critical. Such bearings may be useful in the isolation of structures containing very sensitive equipment. However, for the majority of applications of elastomeric bearings some form of additional energy dissipation is desirable. Significant developments in this direction have been the lead-rubber bearing and the high damping rubber bearing.

Lead-rubber bearings (LRB) are constructed of low damping natural rubber with a pre-formed central hole. A lead core is press-fitted in the hole (Figure 5). The lead core deforms in almost pure shear, yields at low level of stress (8 MPa in shear at normal temperature) and produces hysteretic behavior which is usually stable over a number of cycles. Unlike mild steel, lead recrystallizes at normal temperature (about 20°C), so that repeated yielding does not cause fatigue. Lead rubber bearings exhibit characteristic strength which ensures rigidity under service loads. Lead-rubber bearings were developed in New Zealand and found several applications in New Zealand, Japan and the United States.

High damping rubber bearings (HRB) are made of specially compounded rubber which exhibits equivalent damping ratio of about 0.10 to 0.15 of critical. Originally developed in the United Kingdom, this material found its first application in the first isolated structure in the U.S., the Foothill Communities Center in California (completed in 1986). Bridgestone in Japan subsequently developed a different high damping rubber compound which found several recent applications in Japan.

A comparison of the behavior of elastomeric bearings is provided in Figure 6 which shows test results on full scale natural rubber (NRB), lead-rubber (LRB) and high damping rubber (HRB) bearings. The tests were conducted by the Japanese Central Research Institute of Electric Power Industry (CRIEPI) in a research program for the isolation of Fast Breeder Nuclear Reactors (FBR). The three bearings were designed for gravity load Po=4900kN(500 ton f). They had: NRB, D=1600mm, Σt =219mm; LRB, D=1600mm, Σt =225mm; HRB, D=1420mm, Σt =248mm. All had a shape factor of about 40 (Mazda 1991).

Of interest is to note in Figure 6 the high stiffness of HRB at low shear strain. This property provides the mechanism for rigidity of the isolation system under service loads. It also creates an undesirable effect in earthquakes well below the design level earthquake. If the bearings are designed to shear strain levels of 150% in the design level earthquake, then in frequently occurring moderate level earthquakes the strain will be much less so that the bearings mobilize high stiffness (3 to 4 times more).   Thus, the isolation system may provide insufficient deamplification of the input motion, which may lead to unexpected minor damage.

Another observation to be made in the results of Figure 6 is the difference between the unscragged (or virgin) and scragged (having previoulsly experienced strains) properties of the high damping rubber bearings. It has been assumed in the past that during testing of bearings certain internal structures of rubber are severed so that the scragged properties prevail and become permanent. However, recent studies (Cho 1993, Murota 1994) demonstrated that high damping rubber recovers, either fully or partially, its unscragged (or virgin) properties within a short time interval after testing.

Low damping rubber bearings in combination with yielding mild steel damping devices found several applications in Japan (Kelly 1988). Yielding steel devices were originally applied in New Zealand for the isolation of bridges (Buckle 1990). Wide application of such devices in the seismic isolation of bridges occurred in Italy where over 150 bridges use some form of seismic isolation (Martelli 1993).

Sliding bearings represent another means of providing the basic elements of an isolation system. Flat sliding bearings limit the transmission of force across the isolation interface and thus produce the isolation effect. However, they need to be combined with restoring force devices or otherwise permanent displacements accumulate to unacceptable levels (Constantinou 1991b).   Various means of providing restoring force have been investigated by Constantinou 1993 and Tsopelas 1994. The most practical means of achieving this is by the use of a spherical sliding surface as in the Friction Pendulum System (FPS) bearings (Figure 7).

FPS bearings consist of an articulated slider on a spherical surface which is faced with a polished stainless steel overlay. The slider is faced with bearing material, typically a self-lubricating high bearing capacity composite. Restoring force is generated by the rising of the structure along the spherical surface, while energy is dissipated by friction. The lateral force at the bearing, F, is related to the bearing displacement, u, by (Zayas 1987, Mokha 1991, Constantinou 1993).

$$F = \frac{W}{R} u + \mu W \, sgn(\dot{u}) \tag{5}$$

where W=weight carried by the bearing, R=radius of curvature of the spherical surface, and μ=coefficient of friction. The force is proportional to the carried weight so that the resultant force from all bearings always develops at the center of mass of the isolated structure. This completely eliminates eccentricities in the isolation system even in cases of uneven mass distribution. The quantity W/R represents the stiffness of the bearing. Therefore, the period of isolation system is given by

$$T = 2\pi \left(\frac{R}{g}\right)^{\frac{1}{2}} \tag{6}$$

which is independent of the mass of the structure.

FPS bearings have been used in the United States for the seismic isolation of two tanks, one apartment building in San Francisco, and the U.S. Court of Appeals building, also in San Francisco. This structure with a floor area of 31500m^2 and weight of 55000 ton f (540 MN) is the largest isolated structure in the U.S. (Soong 1992). Figure 8 shows the force-displacement relation of one the FPS bearings of the U.S. Court of Appeals building. Load on the bearing is 2440 kN, displacement amplitude is 290mm and velocity amplitude is 50 mm/sec. The radius of curvature of the bearing is R=1880mm (period T=2.75 secs). Ten cycles of testing demonstrate stable characteristics with a coefficient of friction equal to about 0.06.

WORLDWIDE APPLICATIONS OF SEISMIC ISOLATION

Directories of seismically isolated structures have been given by Buckle 1990, Fujita 1991, Kelly 1993, and Martelli 1993. Combining the information in these references and using personal knowledge, Table 1 was prepared. It presents a brief worldwide directory of seismically isolated structures. It includes only structures either completed or under construction in late 1993. Furthermore, isolated structures have been constructed or are under construction in Chile, China, Greece, Iceland, Iran, Mexico and South Africa.

REFERENCES

1.   American Association of State Highway and Transportation Officials (1991). "Guide specifications for seismic isolation design." Washington, D.C.

2.   Buckle, I.G. and Mayes, R.L. (1990). "Seismic isolation history, application, and performance--a world view." Earthquake Spectra, 6(2), 161-201.

3.   Chalhoub, M.S. and Kelly, J.M. (1990). "Effect of compressibility on the stiffness of cylindrical base isolation bearings." Int. J. Solids Struct., 26(7), 743-760.

4.   Cho, D.M. and Retamal, E. (1993). "The Los Anglees County Emergency Operations Center on high damping rubber bearings to withstand an earthquake bigger than the big one." Proc. ATC-17-1 Seminar on Seismic Isolation, Passive Energy Dissipation, and Active Control, March, San Francisco, CA.

5.   Constantinou, M.C., Kartoum, A., Reinhorn, A.M. and Bradford, P. (1991a). "Experimental and theoretical study of a sliding isolation system for bridges." Report No. NCEER 91-0027, National Center for Earthquake Engineering Research, Buffalo, NY.

6.   Constantinou, M.C., Mokha, A., and Reinhorn, A.M. (1991b). "Study of sliding bearing and helical-steel-spring isolation system." J. Struct. Engrg., ASCE, 117(4), 1257-1275.

7.   Constantinou, M.C., Fujii, S., Tsopelas, P. and Okamoto, S. (1992a). "University at Buffalo-TAISEI Corporation research project on bridge seismic isolation systems." 3rd Workshop on Bridge Engineering Research in Progress, November, La Jolla, CA.

8.   Constantinou, M.C. and Symans, M.D. (1992b). "Experimental and analytical investigation of seismic response of structures with supplemental fluid viscous dampers." Report No. NCEER-92-0032, National Center for Earthquake Engineering Research, Buffalo, N.Y.

9.   Constantinou, M.C. Kartoum, A., and Kelly, J.M. (1992c). "Analysis of compression of hollow circular elastomeric bearings." Eng. Struct., 14(2), 103-111.

10. Constantinou, M.C., Tsopelas, P., Kim, Y-S. and Okamoto, S. (1993). "NCEER-TAISEI corporation research program on sliding seismic isolation systems for bridges-experimental and analytical study of friction pendulum system (FPS)." Report No. NCEER-93-0020, National Center for Earthquake Engineering Research, Buffalo, N.Y.

11. International Conference of Building Officials (1991). "Uniform building code, earthquake regulations for seismic-isolated structures." Whittier, CA.

12. Kelly, J.M. (1991). "Base isolation in Japan, 1988". Report No. UCB/EERC-88/20, University of California, Berkeley.

13. Kelly, J.M. (1993). "State-of-the-art and state-of-the-practice in base isolation." Proc. ATC-17-1 Seminar on Seismic Isolation, Passive Energy Dissipation, and Active Control, March, San Francisco, CA.

14. Martelli, A., Parducci, A. and Forni, M. (1993). "Innovative seismic design techniques in Italy, Proc. ATC-17-1 Seminar on Seismic Isolation, Passive Energy Dissipation, and Active Control, March, San Francisco, CA.

15. Mazda, T., Moteki, M., Ishida, K., Shiojiri, H. and Fujita, T. (1991). "Test on large-scale seismic isolation elements. Part 2: static characteristics of laminated rubber bearing type." Proc. of SMiRT 11, August 18-23, Tokyo, Japan.

16. Mokha, A., Constantinou, M.C., Reinhorn, A.M., and Zayas, V. (1991). "Experimental study of friction pendulum isolation system." J. Struct. Engng., ASCE, 117(4), 1201-1217.

17. Murota, N., Goda, K., Suzuki, S., Sudo, C. and Suizu, Y. (1994). "Recovery characteristics of dynamic properties of high damping rubber bearings." Proc. 3rd U.S.-Japan Workshop on Earthquake Protective Systems for Bridges, January, Berkeley, CA.

18. Soong, T.T. and Constantinou, M.C. (1992). "Base isolation and active control technology-case studies in the U.S.A." IDNDR International Symposium on Earthquake Disaster Reduction Technology, Building Research Institute, Ministry of Construction, Tsukuba, Japan, December.

19. Tsopelas, P., Okamoto, S., Constantinou, M.C., Ozaki, D. and Fujii, S. (1994). "NCEER-TAISEI corporation research program on sliding seismic isolation systems for bridges - experimental and analytical study of systems consisting of sliding bearings, rubber restoring force devices and fluid dampers. "Report No. 94-0002, National Center for Earthquake Engineering Research, Buffalo, N.Y.

20. Winters, C.W. and Constantinou, M. (1993). "Evaluation of static and response spectrum analysis procedures of SEAOC/UBC for seismic isolated structures." Report No. NCEER-93-0004 National Center for Earthquake Engineering Research, Buffalo, N.Y.

21. Zayas, V., Low, S.S. and Mahin, S.A. (1987). "The FPS earthquake resisting system, experimental report." Report No. UCB/EERC-87-01, University of California, Berkeley.

Table 1. Brief Worldwide Directory of Isolated Structures

Country	Type of Structure	Number of Structures	Type of Isolation Systems
Former Soviet Union	Buildings	over 200	Sliding Bearings Rocking Columns Pile-in-sleeve systems
	Bridges	N.A.	N.A.
	Other structures	N.A.	N.A.
France	Buildings	6	Rubber Bearings
	Bridges	N.A.	N.A.
	Other structures	2 Nuclear Power Plants	Neoprene Bearings
Italy	Buildings	9 + several apartment houses of the Italian Navy	High Damping Rubber Bearings Neoprene Bearings
	Bridges	156 (total length = 150 km)	Sliding Bearings Rubber Bearings Lead-rubber Bearings Various Hysteretic Damping Devices
	Other structures	---	---
Japan	Buildings	67	Rubber Bearings & Energy Dissipators Lead-rubber Bearings High Damping Rubber Bearings Sliding Bearings
	Bridges	over 100 partially isolated	Sliding Bearings Lead-rubber Bearings High Damping Rubber Bearings
	Other structures	Radar Tower	Sliding Bearings

Table 1 Brief Worldwide Directory of Isolated Structures (continued)

Country	Type of Structure	Number of Structures	Type of Isolation Systems
New Zealand	Buildings	6	Lead-Rubber Bearings Pile-in-sleeve Systems Lead-rubber & Sliding Bearings
	Bridges	37	Lead-rubber Bearings Various Energy Dissipating Devices
	Other structures	Industrial Chimney	Rocking Foundation
United States	Buildings	24	Lead-Rubber Bearings High Damping Rubber Bearings Friction Pendulum System Springs & Viscodampers
	Bridges	54 (total length = 11 km	Lead-rubber Bearings Sliding Bearings
	Other structures	2 Tanks Heavy Equipment	Friction Pendulum System Lead-rubber Bearings High Damping Bearings Low Damping Bearings

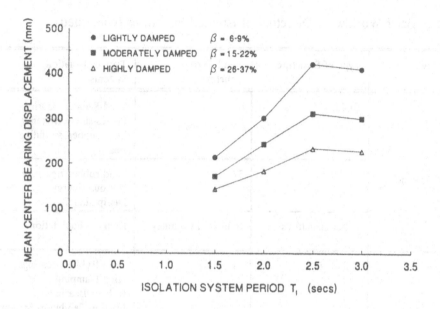

Figure 1   Center Bearing Displacement (Mean of Nine Analyses) in 8-story Structure with Hysteretic Isolation System. Seismic Input Representative of Seismic Zone 4, Soil type S_2 in U.S.

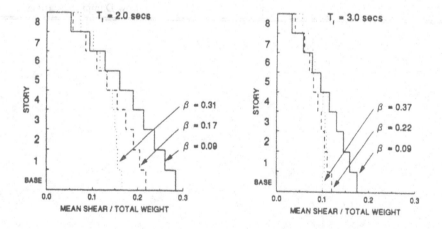

Figure 2   Distribution of Shear Force (Mean of Nine Analyses) with Height in 8-story Structure with Hysteretic Isolation System. Seismic Input Representative of Seismic Zone 4, Soil Type S_2 in U.S.

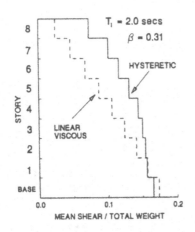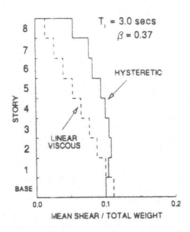

Figure 3   Comparison of Distribution of Shear Force with Height in 8-story
             Structure with Hysteretic and Linear Viscous Isolation System. Seismic
             Input Representative of Seismic Zone 4, Soil Type S_2 in U.S.

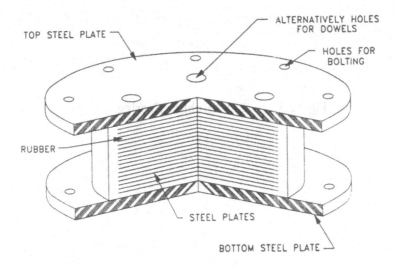

Figure 4   Construction of Elastomeric Bearing

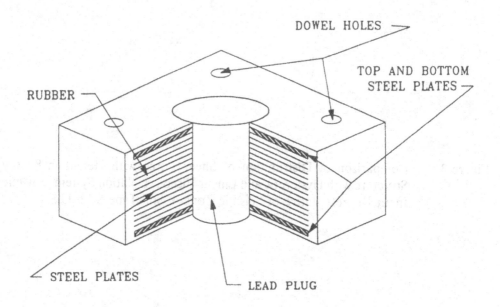

Figure 5   Lead-Rubber Bearing

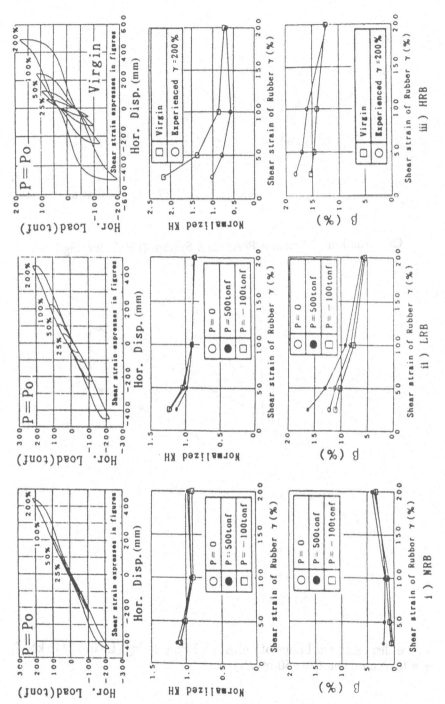

Figure 6   Force-displacement Loops , Horizontal Stiffness and Damping Ratio of Full-scale NRB, LRB and HRB Bearings (figure provided by Dr. Katsuhiko Ishida, Abiko Research Laboratory, CRIEPI, Japan)

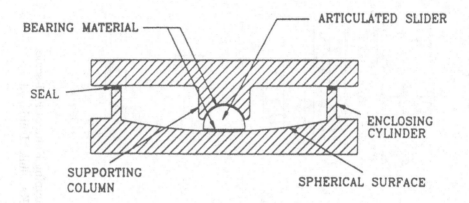

Figure 7   Construction of Friction Pendulum System (FPS) Bearing

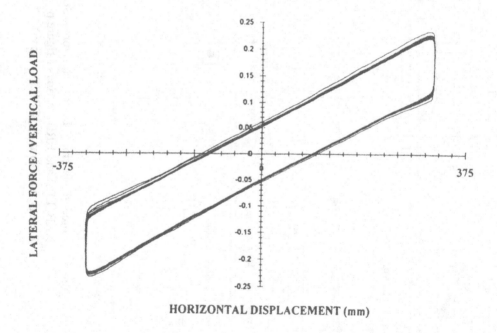

Figure 8   Force-displacement Loops of Full-scale FPS Bearing (Load = 2440 kN, R = 1880 mm, velocity = 50 mm/s)

CHAPTER III

PROPERTIES OF SLIDING BEARINGS:
THEORY AND EXPERIMENTS

M.C. Constantinou
State University of New York at Buffalo, Buffalo, NY, USA

ABSTRACT

Sliding seismic isolation systems may consist of a variety of components, of which the most important one is the sliding bearing. Typically, sliding bearings consist of interfaces made of PTFE or PTFE-based composites and highly polished stainless steel. The properties of these interfaces are described.

INTRODUCTION

Sliding isolation systems found application in a large number of structures in the former Soviet Union (Eisenberg 1992), in a large number of bridges in Italy (Martelli 1993) and a number of buildings, bridges and other structures in Japan and the United States (Fujita 1991, Soong 1992). Little is known about the isolated structures in the former Soviet Union, except that several structures incorporate bearings with PTFE (polytetrafluorethylene or Teflon) in contact with polished stainless steel for the sliding interface. In Japan, sliding bearings consisted of filled (typically glass) PTFE in contact with polished stainless steel. In the United States, sliding bearings consisted of either PTFE or high load capacity composites with PTFE as solid lubricant. In Italy, most applications consisted of lubricated PTFE-polished stainless steel interfaces.

The use of lubrication in the sliding bearings used in Italy reflects the European practice in bridge engineering. Typically in Europe, sliding bearings for accommodating thermal expansion, creep and shrinkage of bridges are lubricated. Lubrication results in very low friction coefficient (approximately 0.01 to 0.02 under dynamic conditions). With such low values of friction, energy dissipation is very low and the isolation system requires some form of energy dissipating mechanism to be activated under seismic conditions. Medeot, 1991 described a variety of such mechanisms which were used in Italy. They include mechanisms based on the yielding of mild steel and fluid damping.

Sliding bearings have also been combined with elastomeric bearings in some applications of seismic isolation in the U.S. and New Zealand. The sliding bearings were used either for providing additional energy dissipation or at wall locations where high compression forces occurred.

The frictional properties of sliding bearings are affected by a number of parameters such as composition and condition of the sliding interface, bearing pressure, velocity of sliding, contamination and temperature. The physical interpretation and quantitative description of these effects is of foremost importance in the design of sliding seismically isolated structures.

FRICTIONAL PROPERTIES OF PTFE AND PTFE-BASED COMPOSITES IN CONTACT WITH POLISHED STAINLESS STEEL

Specifications for PTFE bearings in bridge applications restrict bearing pressures to values below 45 MPa. In the United States the use of unfilled, filled and woven PTFE for pressures up to of 24 MPa, if recessed, is allowed. In Canada and the U.K., bearing pressures of up to 45 MPa on unfilled and filled PTFE are allowed. These limits on pressure depend on the load capacity and wear characteristics of PTFE.

Self-lubricating composite materials are significantly more expensive than PTFE and are typically used in applications with very demanding performance and quality control requirements. Examples include bearings in military and commercial aircraft, satellites, actuators, cranes, hoists, offshore oil platform bridges and FPS bearings (Zayas 1987, Mokha 1991a, Constantinou 1993).

Composite materials in seismic isolation bearings found so far application only in the Friction Pendulum System (FPS) bearings. Over 300 such bearings have manufactured, tested and installed in isolated structures in the U.S. These materials proved to be of exceptional performance with the following characteristics: load capacity of up to 275 MPa pressure, stable frictional properties, very low wear rate, chemical stability and insensitivity to significant temperature changes.

The frictional properties of unlubricated PTFE-stainless steel interfaces under conditions of interest in seismic isolation have been studied by Mokha 1988, 1991b and Constantinou, 1990. In general, we should distinguish between the static (or breakaway) coefficient of friction, μ_B, and the sliding coefficient of friction, μ_s, of these interfaces. The breakaway coefficient of friction is defined as the ratio of shear force to normal force at first movement under extremely slow (i.e. static) conditions of motion. The sliding coefficient of friction is defined as the ratio of shear force to normal force during sliding.

Figure 1 shows the dependency on sliding velocity and pressure of the sliding coefficient of friction of unfilled PTFE in contact with mirror finish polished stainless steel at normal temperature (20° C). Constantinou, 1990 proposed that

$$\mu_s = f_{max} - (f_{max} - f_{min})e^{-a|\dot{u}|} \tag{1}$$

where $\dot{u}$ = velocity of sliding, f_{max} = coefficient of sliding friction at high velocity of sliding, f_{min} = coefficient of sliding friction at essentially zero velocity of sliding, and a = coefficient controlling the dependency of friction on velocity of sliding. Constantinou, 1990 suggests that parameters f_{max}, f_{min} and a are functions of bearing pressure, surface roughness of stainless steel and composition of PTFE (unfilled or filled). Furthermore, temperature has an effect. In general, parameters f_{max}, f_{min}, and a are determined experimentally. Table 1 presents values of these parameters from the tests of Mokha, 1988.

Mokha, 1988 and Constantinou, 1990 found that the breakaway coefficient of friction is, in generally, larger than f_{min} but always less than f_{max}. Values of μ_B are given in Table 1. This fact indicates that the existence of μ_B does not have any adverse effect of the response of seismic sliding isolated structures. This, of course, is valid provided that the value of μ_B does not increase with dwell of load or other effects.

The frictional behavior of self-lubricating composites (containing PTFE) in contact with polished stainless steel follows the basic behavior described for unfilled and filled PTFE. For example, Figure 2 depicts the coefficient of sliding friction of the composite used in the FPS bearings. Interestingly, in this interface, the values of f_{max}, f_{min} and μ_B were found to approach each other at high bearing pressures.

Typically in this interface, $\mu_B \approx f_{min}$ and both are slightly less than f_{max} at pressures larger than 100 MPa.

In summary, the following interesting conclusions may be derived for the friction of PTFE and PTFE-based materials in contact with polished stainless steel:

(1)   The coefficient of sliding friction reduces with increasing bearing pressure at all velocities of sliding. However, the rate of reduction appears to depend on the velocity of sliding. It is greatest at high velocity of sliding.

(2)   The effect of bearing pressure on the sliding friction diminishes at a value of pressure which depends on the type of PTFE and on the velocity of sliding. This limit of pressure is generally larger than about 35 MPa.

(3)   For sheet type PTFE (unfilled or filled), the sliding friction at pressures beyond the approximate limit of 35 MPa appears to be unaffected by the composition of PTFE (with or without fillers) and the surface roughness of stainless steel provided that the roughness is less than about 0.07 μm of arithmetic average (Mokha 1988). For PTFE-based composites (see Figure 2), a similar behavior occurs at higher pressure, approximately 100 MPa.

EFFECT OF TIME AND TEMPERATURE ON THE FRICTION PROPERTIES OF SLIDING BEARINGS

Time may have the following effects:

(1) Alter the properties of the sliding interface by
        (a) Contamination
        (b) Corrosion of stainless steel
        (c) Chemical deterioration of PTFE or PTFE-based composite

(2)   Increase of adhesion or true contact area due to viscoelastic effects which may cause an increase in the breakaway coefficient of friction.

Contamination

PTFE is comparatively soft and capable of absorbing light contamination without effect on the coefficient of friction (Campbell 1987). However, heavy contamination causes an increase of friction during the initial cycle of movement (Campbell 1987, Mokha 1991). Campbell, 1989 reported also significant effect of light contamination on the friction of lubricated bearings. Campbell, 1989 found that contamination of bearings in service in a dust-laden environment is unlikely, even when the bearings are unprotected and with a significant eccentricity of loading.

However, it is a prudent practice to provide sealing to sliding bearings and not to disassemble bearings on site.

Chemical Stability of PTFE and PTFE-based Composites

PTFE has unique inertness as a result of the super-strong inter-atomic bonds which form PTFE. Conclusive tests on PTFE samples exposed for over fifteen years to practically all climatic conditions confirm the long-term chemically stability and weather resistance of PTFE (du Pont, 1981).

PTFE-based composites are used in aerospace and military applications so that they qualify under a number of U.S. Military Specifications. Particularly, the composite material used in the FPS bearings (Figure 2) is rated by these specifications as chemically stable and inert, without aging effects.

Corrosion of Stainless Steel

The stainless steel used in the sliding bearings is typically polished (to mirror finish) austenitic, type 304 or 316. Of the four groups of stainless steels (austenitic, ferritic, martensitic and precipitation-hardening), the austenitic group is the most resistant to corrosion. Particularly, the 316 type with high content of chromium and nickel together with a small amount of molymbenum is virtually impervious to rusting from almost any outdoor exposure.

The excellent corrosion resistance of stainless steels is well known. A collection of data on the corrosion resistance of stainless steel is included in the Metals Handbook on Corrosion of ASM International (ASM, 1987). The handbook presents data on the corrosion of austenitic stainless steel after exposure of

(a)   23 years in industrial and marine environment in New York City, and
(b)   15 years in marine environment at 250m from the ocean in North Carolina.

In both studies the conditions of exposure were severe, that is the specimens were unprotected. The condition of the surface of the stainless steel is described as free from rust stains. In particular, of all the types of the studied stainless steel, the 316 type had the best performance with the passivation film developed over the least area. This passivation film could be easily removed (that is removed by hand) to reveal the bright finish.

Further explicit information on the finish condition of aged highly polished stainless steel has been recently obtained by this author through the inspection of the inventory of old stainless steel parts of a Buffalo area manufacturer. Among several inspected parts, a number was stored outside, completely unprotected for 10 years. The finish of all samples was in perfect condition. Furthermore, seven inspected parts were in storage for 34 years in an indoor environment equivalent to sealed bearings. All parts were in excellent condition with perfectly reflective finish.

Effect of Dwell of Load

The possible increase of the breakaway coefficient of friction with time (or dwell) of loading has been investigated by some researchers. Mokha, 1991b studied the breakaway friction of unfilled Teflon in contact with polished stainless steel at a pressure of 13.8 MPa and dwell of load of 1/2 hour and 594 days. Furthermore, Mokha, 1991b studied the breakaway friction of woven PTFE at a pressure of 45 MPa and dwell of load of 1/2 and 7 days. The results demonstrate that the dwell of load has no or insignificant effect on the breakaway friction. Similar results have been reported by Hakenjos, 1974.

Studies of Paynter, 1973 on unfilled PTFE at pressure of 24 MPa show that the breakaway friction is affected by dwell of load up to 24 hours. Beyond this time limit, dwell has no significant effect.

These studies point to one common observation. Dwell of load may have effects which are limited to times from a few minutes to a few hours and not several years. The mechanism responsible for this phenomenon is identified in the next section.

Effect of Temperature

Campbell, 1991 reported data on the friction of PTFE in contact with polished stainless steel under conditions of very low sliding velocity and temperature in the range of 20° to -25°C. Unfilled and filled PTFE appear unaffected by temperature above 0°C but they show a marked increase in friction at temperatures below -10°C. It appears that this behavior is related to the glass transition temperature of PTFE which is -13°C (Gardos 1982).

It is known that the increase of friction with decreasing temperature in the PTFE-based composites is much less than in unfilled PTFE. This actually represents a leading reason for the use of these materials in the ultra-cold conditions of aerospace applications (Gardos 1982). Specific data on the frictional properties of one such material have been recently obtained from testing of full-size FPS bearings. Within the range of 30° to -30°C, the coefficient of friction at high velocity of sliding (f_{max}) increased approximately by 0.01 per 25°C reduction in temperature.

PHYSICAL INTERPRETATION OF FRICTION IN SLIDING BEARINGS

It is generally accepted that the friction between clean surfaces is mainly the result of **adhesion** (Bowden and Tabor, 1964). That is , the friction force is given by the product of real area of contact and the shear strength of the junctions.

At low velocity of sliding , PTFE is transferred to the smooth substrate (stainless steel) as a very thin, highly oriented film. The film is 10 to 40 nm thick and oriented in the sliding direction. The shear strength, s, of the interface is very low. The friction force during sliding is

$$F = sA \tag{2}$$

where A is the true area of contact

The true contact area depends on the normal load. At low load (low average bearing pressure) the contact is elastic so that

$$A = kW^{\eta} \tag{3}$$

where k = coefficient, W = normal load, η = coefficient in the range of 2/3 to 1. The coefficient 2/3 is derived when considering that the PTFE surface is covered with spherical asperities. Values of η larger than 2/3 are derived by considering more complicated surface asperity distribution (Bowden and Tabor, 1964). At high normal load the true area of contact is

$$A = \frac{W}{p_y} \tag{4}$$

where p_y = yield strength of PTFE.

The coefficient of sliding friction is, by definition,

$$\mu_s = \frac{F}{W} \tag{5}$$

so that

$$\mu_s = \frac{sK}{W^{1-\eta}} \tag{6}$$

under elastic conditions, and

$$\mu_s = \frac{s}{p_y} \tag{7}$$

under plastic conditions.

Equations (6) and (7) explain the dependency of the coefficient of sliding friction on bearing pressure. At low pressure the contact is elastic so that μ_s decreases with increasing W (note that $1-\eta \geq 0$). At high pressure the contact is plastic and μ_s is independent of normal load. The compressive yield strength of PTFE is about 15 MPa but the pseudo-yield strength of confined sheet PTFE is certainly much higher. This is consistent with the observed limit of about 35 MPa beyond which the coefficient of sliding friction is independent of normal load.

As the velocity of sliding is increased the force needed to shear the film at the sliding interface increases. This phenomenon is primarily a result of the viscoelasitc nature of PTFE. At higher velocities a stage is reached where the shear force exceeds the strength of the boundaries between crystals. This higher fiction is then accompanied by the transfer of large fragments of PTFE (Mokha, 1988).

In summary, the following may be concluded:

(1)   The dependency of the friction on the true area of contact is important at low bearing pressures, that is, pressures well below the pseudo-yield strength of PTFE, p_y. Under these conditions, it is expected that dwell of load will result in an increase in the true area of contact and thus, increase of the coefficient of friction. The process by which this happens is the creep of PTFE. The process depends on the viscoelastic properties of PTFE. Therefore, there is a limit of time beyond which the true area of contact does not change and friction is not affected. The experimental results discussed in the previous section suggest that this time varies from a few minutes to a few hours.

(2)   At high bearing pressures, as those in bearings utilizing composites (e.g. FPS bearings), the true area of contact is constant and creep effects are insignificant. Accordingly, dwell of load should have no effect.

(3)   The dependency of the coefficient of friction on the velocity of sliding is a result of the viscoelastic behavior of PTFE. In essence, the shear strength of the interface is dependent on the rate of loading.

REFERENCES

1. ASM International (1987). Metals Handbook, Vol. 13 on Corrosion, Metals Part, Ohio.

2. Bowden, F.P. and Tabor, D. (1964). The friction and lubrication of solids. Part II, Oxford University Press.

3. Campbell, T.I. and Kong, W.L. (1987) "TFE sliding surfaces in bridge bearings." Report ME-87-06, Ontario Ministry of Transportation and Communications, Ontario, July.

4. Campbell, T.I. and Fatemi, M.J., (1989) "Further laboratory studies of friction in TFE slide surface of a bridge bearing." Report MAT-89-06, Ministry of Transportation and Communications, Ontario, October.

5. Campbell, T.I. Pucchio, J.B., Roeder, C.W. and Stanton, J.F. (1991). "Frictional characteristics of PTFE used in sliding surfaces of bridge bearings." Proc. 3rd World Congress on Joints Sealing and Bearing Systems for Concrete Structures, Vol. 2, 847-870, Toronto, October.

6. Constantinou, M.C., Mokha, A.M. and Reinhorn, A. (1990). "Teflon bearings in base isolation. II: modeling." J. Struct. Engng., ASCE, 116(2), 455-474.

7. Constantinou, M.C., Tsopelas, P., Kim, Y-S. and Okamoto, S. (19993). "NCEER-TAISEI corporation research program on sliding seismic isolation systems for bridges - experimental and analystical study of friction pendulum system (FPS)." Report No. NCEER-93-0020, National Center for Earthquake Engineering Research, Buffalo, N.Y.

8. E.I. du Pont de Nemours & Co. (1981). "Teflon: Mechanical design data." Wilmington, DE.

9. Eisenberg, J.M., Melentyev, A.M., Smirnov, V.I. and Nemykin, A.N. (1992). "Applications of seismic isolation in the USSR." Proc. 10th WCEE, Vol. 4, 2039-2046, Madrid.

10. Fujita, T. editor (1991). "Isolation and response control of nuclear and non-nuclear structures." Special Issue for the Exhibition of SMiRT 11, August 19-23, Tokyo, Japan.

11. Gardos, M.N. (1982). "Self-lubricating composites for extreme environmental applications." Tribology International, October, 273-283.

12. Hakenjos, V. and Richter, K. (1974). "Experiments with sliding supports", in Lager and Bauwesen, edited by H. Eggert et al., Wilhelm Ernst und Sohn, Berlin.

13. Martelli, A., Parducci, A. and Forni, M. (1993). "Innovative seismic design techniques in Italy." Proc. ATC-17-1 Seminar on Seismic Isolation, Passive Energy Dissipation, and Active Control, March, San Francisco, CA.

14. Medeot, R. (1991). "The evolution of aseismic devices for bridges in Italy." 3rd World Congress on Joint Sealing and Bearing Systems for Concrete Structures, Vol. 2, 1295-1320, Toronto.

15. Mokha, A., Constantinou, M.C. and Reinhorn, A.M. (1988). "Teflon bearings in aseismic base isolation: experimental studies and mathematical modeling." Report No. NCEER-88-0038, National Center for Earthquake Engineering Research, Buffalo, NY.

16. Mokha, A. Constantinou, M.C., Reinhorn, A.M., and Zayas, V. (1991a). "Experimental study of friction pendulum isolation system." J. Struct. Engng., ASCE, 117(4), 1201-1217.

17. Mokha, A., Constantinou, M.C. and Reinhorn, A.M. (1991b). "Further results on the frictional properties of Teflon bearings." J. Struct. Engng., ASCE 117(2), 622-626.

18. Paynter, F.R. (9173). "Investigation of friction in PTFE bridge bearings." Civil Engineer in South Africa, 15(8), Aug., 209-217.

19. Soong, T.T. and Constantinou, M.C. (1992). "Base isolation and active control technology-case studies in the U.S.A." IDNDR International Symposium on Earthquake Disaster Reduction Technology, Building Research Institute, Ministry of Construction, Tsukuba, Japan, December.

20. Zayas, V., Low, S.S. and Mahin, S.A. (1987). "The FPS earthquake resisting system, experimental report." Report No. UCB/EERC-87-01, University of California, Berkeley.

Table 1:   Frictional Properties of PTFE in Contact with Polished Stainless Steel

Type of PTFE	Pressure MPa	Sliding direction	f_{max} (%)	f_{min} (%)	μ_B (%)	μ_B/f_{max}
UF	6.9	P	11.93	2.66	5.85	0.49
UF	13.8	P	8.70	1.75	4.03	0.46
UF	20.7	P	7.03	1.51	5.59	0.79
UF	44.9	P	5.72	0.87	3.74	0.65
15GF	6.9	P	14.61	4.01	8.42	0.58
15GF	13.8	P	10.08	4.28	6.00	0.59
15GF	20.7	P	8.49	4.32	5.62	0.66
15GF	44.9	P	5.27	2.15	4.73	0.90
25GF	6.9	P	13.20	5.54	7.76	0.59
25GF	13.8	P	11.20	4.87	6.82	0.61
25GF	20.7	P	9.60	4.40	6.60	0.69
25GF	44.9	P	5.89	3.19	5.74	0.97
UF	6.9	T	14.20	2.39	7.17	0.50
UF	13.8	T	10.50	1.72	7.57	0.72
UF	20.7	T	8.20	2.90	4.35	0.53
UF	44.9	T	5.50	1.11	3.55	0.64

Note: UF = unfilled PTFE; 15GF = glass-filled PTFE at 15%; 25GF = glass-filled PTFE at 25%; P = sliding parallel to lay (roughness Ra = 0.04 μm,, AA or CLA) T = sliding perpendicular to lay (roughness R_a = 0.07 μm, AA or CLA)

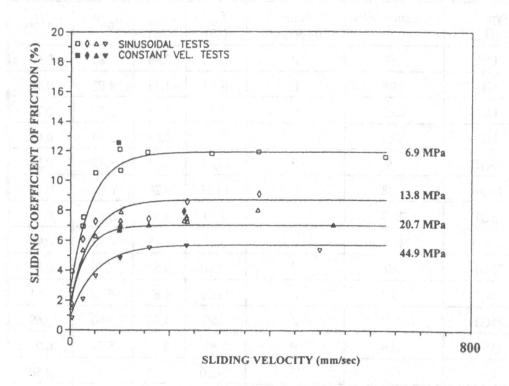

Figure 1 -   Variation of Sliding Coefficient of Friction with Velocity and Pressure
of Unfilled PTFE Sliding Against Stainless Steel Polished to 0.04 $\mu m R_a$

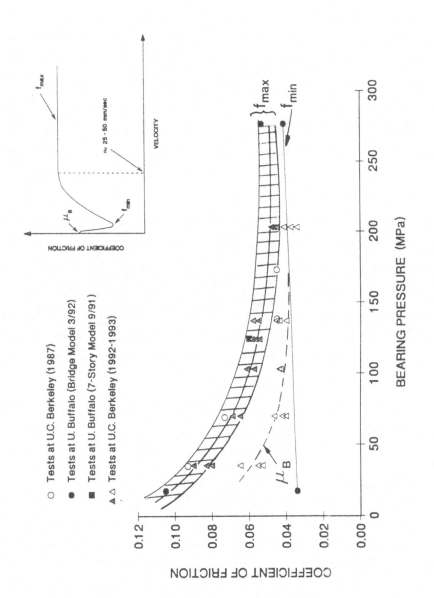

Figure 2 - Frictional Properties of PTFE-based Composite in FPS Bearings

CHAPTER VI

DESIGN AND APPLICATIONS OF SLIDING BEARINGS

M.C. Constantinou

State University of New York at Buffalo, Buffalo, NY, USA

ABSTRACT

The design of sliding isolation systems and relevant applications are presented. Sliding isolation systems which found application are:

(1)   EDF system consisting of leaded bronze-stainless steel sliding bearings without restoring force,

(2)   TASS system consisting of PTFE-elastomeric sliding bearings and rubber restoring force devices,

(3)   Spherical sliding or Friction Pendulum System (FPS),

(4)   Lubricated PTFE sliding bearings with additional energy dissipating devices, and

(5)   Sliding bearings with restoring force devices for bridge applications.

EDF SYSTEM

Spie-Batignolles Batiment Travaux Publics (SBTP) and Electricite de France (EDF) of France developed a sliding seismic isolation system for application in nuclear power plants. Development of the system started in the 1970's and resulted in an application in the Koeberg nuclear power plant in South Africa. Another application at the Karun plant in Iran proceeded to the point of manufacturing of the bearings but construction was halted following the revolution in Iran (Plichon 1980, Pavot 1979, Gueraud 1985).

The isolated plant in South Africa is 150m by 100m in plan, consists of two 900 MWe units and weighs in excess of 100,000 metric tons (980 MN). About 2000 bearings on a total of 600 pedestals support the entire nuclear island. In terms of plan dimensions, weight and number of bearings, this 15 year old plant represents by far the largest isolated structure.

Figure 1 depicts the EDF system configuration and typical response to seismic excitation. Each bearing consists of a lower elastomeric part (typically neoprene) and a top sliding interface consisting of leaded bronze in contact with stainless steel. The bronze plate has grooves on the surface in order to collect debris in the wear process and prevent it from entering in the sliding interface. Under these conditions the interface exhibits essentially Coulomb type friction with a friction coefficient in the range of 0.18 to 0.22 at pressures of 2 to 15 MPa and at all sliding velocities. The lead acts as solid lubricant to ensure stable frictional properties.

The supporting elastomeric bearing part is typically designed to provide a frequency of about 1 Hz (actually 0.9 Hz in the Koeberg plant) to the isolated structure when sliding does not occur. For seismic excitation below a certain threshold (peak ground acceleration below about 0.2 g), sliding does not occur and the isolated structure responds with only distortion in the elastomeric part. In stronger excitation sliding occurs (see Figure 1), which effectively limits the force transmitted to the superstructure to about 0.2 times the weight (coefficient of friction x weight). Thus, a standardized plant design may be used in areas of significantly different seismicity without the necessity for extensive analyses and qualification testing.

The EDF system lacks restoring force capability so that permanent displacements occur (see Figure 1), which may accumulate to large values in sequential earthquakes. This may be a problem in applications in which the Operating Basis Earthquake (OBE) is strong enough to induce sliding of the bearings. Thus in earthquakes which are expected to occur frequently during the lifetime of the structure, moderate sliding will occur and permanent displacements will accumulate to unacceptable values. This has not been a problem with the Koeberg plant since the OBE had a peak acceleration below 0.2g which could not induce sliding of the bearings.

Recent seismic isolation design specifications in the U.S. (ICBO 1991, AASHTO 1991) penalize isolation systems which lack restoring force capability. Based on the possibility for accumulation of large permanent displacements, these specifications require that such isolation systems be designed for a displacement at least equal to three times the displacement calculated for the design earthquake.

TASS SYSTEM

The TASS system was developed by Taisei Corporation, Japan. It has been applied to three structures in Japan. These are the Taisei Technology Research Center building in Yokohama, a 4-story, 1173m^2 floor area structure, the JAPCO Atagawa building in Shizuoka, a small single story indoor pool structure and the Toho Gas Center in Mie, a 3-story, 1800m^2 floor area building.

The TASS system consists of PTFE-elastomeric bearings to support the weight and provide the isolation mechanism and chloroprene rubber (neoprene) springs to provide weak restoring force and displacement restraint. Figure 2 illustrates the configuration of the system and its response to seismic excitation. (Kawamura 1988).

The sliding interface of the PTFE-elastomeric bearings consists of glass-filled PTFE in contact with polished stainless steel and at pressure of about 7 MPa. The coefficient of friction at high velocity of sliding (f_{max}) is about 0.14, whereas at very low velocity of sliding (f_{min}) is about 0.04 (Mokha 1988). Denoting as K_1 the stiffness of the elastomeric part of the PTFE-elastomeric bearings and as K_2 the stiffness of the neoprene springs, the idealized force-displacement relation of the TASS system takes the form of Figure 3. The combined stiffness $K_1 + K_2$ is selected so that the rigid body period of the isolated structure prior to sliding, T_1, is between 1 and 2 secs:

$$T_1 = 2\pi \left[\frac{W}{g(K_1 + K_2)} \right]^{1/2} \qquad (1)$$

where W=weight. Thus, sliding does not occur as long as the isolation system displacement is below the value Y

$$Y = \frac{f_{min}g}{4\pi^2} T_1^2 \tag{2}$$

For period T_1 in the range 1 to 2s, Y equals 10 to 40 mm. Therefore, for weak seismic excitation the isolation system behaves as a lightly damped elastomeric system which provides significant de-amplification of the input motion.

In strong seismic excitation, sliding occurs and the low stiffness (K_2) of the neoprene springs is activated. This stiffness is typically selected as to give a period of about 5 secs. Under these conditions, the maximum force transmitted through the isolation interface is

$$V_b = f_{max}W + K_2(d-Y) \tag{3}$$

or

$$\frac{V_b}{W} = f_{max} + 4\pi^2 \frac{d-Y}{gT_2^2} \tag{4}$$

where T_2 is the period of approximately 5 secs, and d=maximum isolation system displacement.

Experimental studies (including triaxial excitation on a shake table) and analyses demonstrated that for Japan the maximum displacement d is less than 300mm and typically about 200mm. Thus, the maximum isolation system shear force (equation 4) is less than 0.19W and typically about 0.15W. Isolated structures in Japan are designed for a base shear coefficient of 0.15 to 0.2. The value of 0.2 is the minimum allowed by the code based on the building period. Reduction to 0.15 is allowed after analytical and/or experimental evidence is provided.

The configuration of the TASS system allows for ease in design. The characteristic strength, initial stiffness and post-sliding stiffness of the system are controlled by different, non-interacting elements. Proper selection of these characteristics may lead to significant de-amplification of the input motion in both weak and strong seismic excitation.

FRICTION PENDULUM SYSTEM (FPS)

FPS bearings were developed by Earthquake Protection Systems, San Francisco and since 1987 extensively studied at U.C. Berkeley and the Univeristy at Buffalo (Zayas 1987, Mokha 1991, Zayas, Constantinou 1993). FPS bearings combine the basic elements of an isolation system (flexibility and energy dissipation capability) in a compact, low profile, all steel unit.

FPS bearings found application in a water tank in San Francisco, an ammonia tank in Calvert City, Kentucky, the Hawley Apartments building, a 4-story $1900m^2$ floor area building in San Francisco, and the U.S. Court of Appeals, also in San Francisco. This massive 540 MN, $31500m^2$ floor area building is the largest isolated structure in the United States (Soong 1992, Amin 1993).

A cross section view of an FPS bearing is shown in Figure 4. The bearing consists of a spherical sliding surface and an articulated slider which is faced with a high pressure capacity composite bearing material.

The basic principle of operation of the FPS bearing is illustrated in Figure 5. The motion of a structure supported by these bearings is identical to that of pendulum motion with the additional beneficial effect of friction at the sliding interface. The force needed to produce displacement of the FPS bearing consist of the combination of restoring force during the induced rising of the structure along the spherical surface and of friction force at the sliding interface. The derivation of the force-displacement relation is based on Figure 6. The FPS bearing is considered in its deformed position under the action of a lateral force F. The horizontal and vertical components of displacement are respectively given by

$$u = R\sin\theta \qquad (5)$$

$$v = R(1 - \cos\theta) \qquad (6)$$

where R is the radius of curvature of the spherical sliding surface. From equilibrium of the bearing in the vertical and horizontal directions it is derived that

$$F = W\tan\theta + \frac{F_f}{\cos\theta} \qquad (7)$$

where W is the weight carried by the bearing and F_f is the friction force at the sliding interface. The quantity $W\tan\theta$ represents the restoring force in its most general form (θ may be large).

The stiffness of the bearing is derived by dividing the restoring force by the displacement u

$$K = \frac{F}{u} = \frac{W}{R\cos\theta} \qquad (8)$$

Accordingly, the force needed to induce a displacement u in the horizontal direction is given in the general case of large θ by

$$F = \frac{W}{R\cos\theta}u + \frac{F_f}{\cos\theta} \tag{9}$$

For small values of angle θ, $\cos\theta \approx 1$ and Equations (8) and (9) take the linearized form

$$K = \frac{W}{R} \tag{10}$$

$$F = \frac{W}{R}u + \mu W sgn(\dot{u}) \tag{11}$$

in which now the friction force F_f has been replaced by the product of the coefficient of friction μ and weight W. Furthermore, $\dot{u}$ represents the horizontal component of velocity. Equations 10 and 11 are valid for all practical purposes. FPS bearings are typically designed for displacement $u < 0.2R$, so that the error of linearizing Equations (8) and (9) is insignificant.

Returning now to Equation (10), we calculate the period of free vibration as

$$T = 2\pi\left(\frac{W}{Kg}\right)^{1/2} = 2\pi\left(\frac{R}{g}\right)^{1/2} \tag{12}$$

The period is independent of the mass of the structure and dependent only on the geometry of the bearing. Thus, the period does not change if the weight of the structure changes or it is different than assumed.

Furthermore, Equation (11) demonstrates that the lateral force is directly proportional to the supported weight. As a result of this significant property the center of lateral rigidity of the isolation system coincides with the center of mass of the structure. This property makes the FPS bearings particularly effective at minimizing adverse torsional motion in asymmetric structures.

As an example of force-displacement relation of FPS bearings, Figure 7 shows recorded loops of isolation system force (four bearings) versus bearing displacement in recent tests at the National Center for Earthquake Engineering Research, Buffalo, N.Y. (Constantinou, 1993). A 158 kN, flexible pier bridge structure was supported by four scaled FPS bearings having R = 558.8 mm. The excitation in this test was the 1940 El Centro, CA earthquake scaled up by a factor of two (to a peak acceleration of 0.68 g). The slope of the force-displacement loop is precisely that given by Equation (10).

FPS bearings, in addition to combining the basic elements of an isolation system in a single and compact unit, have the following unique properties:

(a) Period of vibration which is independent of the supported mass.

(b) Their lateral force is directly proportional to the weight they carry and thus, the isolation system center of rigidity coincides with the center of mass of the supported structure. This property minimizes adverse torsional motions. Shake table tests with uneven distribution of mass (up to 80% of the total mass supported by half of the total number of bearings) resulted in insignificant torsional motion of the tested isolated structure (Zayas 1987, 1993; Mokha 1991)

LUBRICATED PTFE BEARINGS WITH ENERGY DISSIPATING DEVICES

Italian engineers developed a number of sliding isolation systems for bridges. Approximately 150 bridges in Italy employ some form of these systems for seismic isolation. An account of these bridges and some details of the isolation systems have been presented by Medeot, 1991 and Martelli, 1993. Typically, the isolation system consists of lubricated PTFE sliding bearings together with some form of energy dissipating device such as fluid dampers or yielding mild steel elements. One of the most interesting applications is that of the Mortaiolo viaduct on the Livorno-Civitavecchia highway. Constructed in 1990, the viaduct has an isolated total length of 8 km. It consists of isolated continuous sections, each one of which has length of about 426m and it is divided into 10 spans of length equal to either 33m or 45m. Each end of a span is supported by piers (Marioni 1991).

A combination of bearings is used in the Mortailo viaduct as shown in Figure 8. The four end bearings in each 426m long section are multidirectional lubricated PTFE sliding bearings. The two central bearings are also multidirectional lubricated bearings, however equipped with E-shaped mild steel dampers. The bearings in-between the end and central bearings are also equipped with E-shaped steel dampers, however they employ shock transmission units (seismic snappers) which activate the dampers only in motions with velocity exceeding about 1 mm/s. Thus under service loading conditions, each section behaves as a standard bridge which allows thermal expansion about the central fixed bearings. The action of the bearings with shock transmission units is illustrated in Figure 9. Under slow longitudinal motion, as in thermal expansion, the E-shaped dampers do not deform and allow for unrestricted movement. Under seismic excitation, the shock transmission units lock and the E-shaped dampers deform as illustrated in Figure 9.

It should be noted that E-shaped dampers are also installed in the transverse direction, so that isolation is multidirectional (Marioni 1991).

The action of the E-shaped damper is double. First is to provide rigidity against service loads at selected locations. Second is to yield and dissipate energy in seismic excitation. Deformation of the E-shaped damper (see Figure 9, the central leg moves with respect to the two exterior legs) induces constant bending moment over the entire length of the beam. On yielding, plasticization occurs over the entire length of the beam. This extends the low-cycle fatigue life of the device (Ciampi 1991).

An example of behavior of a lubricated bearing with E-shaped dampers is shown in Figure 10. The geometry of the scaled bearing is given in Figure 11. Four of these bearings supported a quarter length scale model bridge deck of weight equal to 140 kN during shake table testing at the Univ. at Buffalo-National Center for Earthquake Engineering Research in 1992 (Constantinou, 1994). The model was subjected to 16 seismic tests and then the bearings were subjected to 25 cycles of ± 50 mm displacement. The displacement at full plasticization was estimated to be about 5 mm, so that the global ductility at 50 mm was about 10. The maximum value of the strain in the dampers was approximately equal to 0.03. The dampers sustained 25 cycles of motion and exhibited stable properties up to the last cycle in which failure occurred. Overall the bearings maintained their properties for a total of about 50 cycles at global ductility exceeding 10 (displacement exceeding 50 mm).

Results obtained in the shake table testing of the quarter scale bridge model are presented in Figure 12. It should be noted that the bearings were designed to fully yield at approximately 0.15 times the carried weight (W) which together with friction of the lubricated sliding interface would give approximately 0.17 W. Testing gave a force of about 0.2 W as a result of higher yield stress of steel.

In one of the tests the model was excited with the Japanese bridge design motion of level 2, ground condition 3 (deep soil) scaled down to 0.3 g (~ 75% of full motion). The bearings performed very well in this long duration motion. Permanent displacements were insignificant. The other test was with the 1952 Taft motion scaled up to 0.6 g. The input consisted of primarily a single strong shock. While the force transmitted to the deck and bridge substructure below was effectively limited to 0.2 W, permanent displacements of over 50% of the bearing displacement capacity were recorded. This rendered the bearings incapable of sustaining the same motion again.

The lack of restoring force in this system represents a drawback which in the current design philosophy in the U.S. is penalized (AASHTO 1991). If the scaled Taft motion represented the design earthquake for an isolated bridge with the characteristics of the tested model, the design displacement should have been 4 x 60 = 240 mm (from test results after extrapolation to prototype scale). However, the

1991 AASHTO would require that the bearings be designed for a displacement capacity of 3 x 240 mm or 720 mm. With such requirement on the displacement capacity the bearings would have been excessively large in comparison to other isolation systems (such as lead-rubber or FPS bearings) which could be easily designed to have an isolation system force of 0.2 W with design displacement not exceeding 200 mm.

Apparently, design philosophies in various countries differ. The U.S. practice would penalize systems without sufficient restoring force. The Japanese practice would allow such systems but it would require re-centering of the structure following the development of significant permanent displacement.

SYSTEMS CONSISTING OF SLIDING BEARINGS AND RESTORING FORCE DEVICES

A number of bridge seismic isolation systems, which consist of flat PTFE sliding bearings and various forms of restoring force devices, have been developed and tested at the University at Buffalo (Constantinou 1991, Constantinou 1994, Tsopelas 1994). The restoring force devices consisted of either unidirectional simple spring-like devices and advanced compressible fluid-damping devices or multidirectional rubber springs. Figure 13 shows two such devices and representative force-discplacement loops.

Very recently, a number of bridges in the United States have been seismically isolated using PTFE sliding bearings with restoring force devices in the form of unidirectional urethane rubber springs. The system has the characteristics of the generic system tested by Constantinou 1991, except that the sliding bearings and restoring force devices have been combined in a single compact unit.

MATHEMATICAL MODELING OF FRICTION IN SLIDING BEARINGS

The friction force which is mobilized at the sliding interface of sliding bearings depends on the normal load, bearing pressure, direction and value of sliding velocity and composition of the sliding interface. Constantinou, 1990 and Mokha, 1993 described a mathematical model to account for these effects on the frictional forces which develop as a result of sliding.

At an instance of time a sliding bearing undergoes displacement components U_x and U_y and velocity components $\dot{U}_x$ and $\dot{U}_y$ along the two orthogonal directions. Furthermore, the interface carries a weight W, and it is subjected to vertical acceleration $\ddot{U}_v$ and additional seismic load P_s due to overturning moments effects. The frictional forces F_x and F_y which develop at the sliding interface are described by

$$F_x = \mu_s W^* Z_x \qquad (13)$$

$$F_y = \mu_s W^* Z_y \qquad (14)$$

where

$$W^* = W(1 + \frac{\ddot{U}_v}{g} + \frac{P_s}{W}) \qquad (15)$$

is the normal load on the sliding interface. Furthermore, Z_x and Z_y are dimensionless variables which are governed by a system of differential equations (Constantinou 1990). These variables account for the conditions of separation and re-attachment of the sliding interface. They take values in the ranges $\pm \cos \theta$ for Z_x and $\pm \sin \theta$ for Z_y, where

$$\theta = \tan^{-1}\left(\frac{\dot{U}_y}{\dot{U}_x}\right) \qquad (16)$$

Angle θ is defined as the angle between axis x and the direction of instantaneous sliding.

The coefficient of sliding friction, μ_s, is described by

$$\mu_s = f_{max} - (f_{max} - f_{min})e^{-a|\dot{U}|} \qquad (17)$$

where

$$\dot{U} = (\dot{U}_x^2 + \dot{U}_y^2)^{\frac{1}{2}} \qquad (18)$$

is the magnitude of the instantaneous velocity. Parameters f_{max}, f_{min} and a are functions of the composition of the sliding interface, surface roughness and bearing pressure as described in the section on Properties of Sliding Bearings: Theory and Experiments. Typically, these parameters may be assigned constant values based on the bearing pressure calculated for the gravity load (weight W divided by bearing area). More accuracy, although of little practical significance, is obtained by describing f_{max} as function of the instantaneous bearing pressure p:

$$p = \frac{W}{A}\left(1 + \frac{\ddot{U}_v}{g} + \frac{P_s}{W}\right)$$ (19)

where A = bearing area. A relation found to approximate very well the experimental data is (Constantinou 1993).

$$f_{max} = f_{max_o} - Df \tanh(\alpha p)$$ (20)

Mokha 1993 conducted tests of PTFE bearings under constant normal load and two-dimensional plane motion. Figure 14 shows the motion imposed in one of the tests. The motion consists of 45mm equal amplitude harmonic components in both orthogonal directions with frequency of 0.35 Hz in the x direction and 0.7 Hz in the y direction (test No. 6). In another slower test (No. 3), the frequencies were 0.08 Hz and 0.16 Hz, respectively. Recorded and analytically determined loops of friction force versus displacement compared very well as shown in Figure 14. These loops were obtained from simultaneous testing of two PTFE bearings under normal load of 48 kN.

In another example, Figure 15 shows the response of the aforementioned quarter length scale bridge model from testing on the University at Buffalo shake table (Constantinou 1993). Four FPS bearings with R = 559mm and f_{max} = 0.104 (composite material under pressure of 17.3 MPa) supported a 140 kN deck on top of flexible piers. The model was excited with the 1952 Taft motion, component N21E in the horizontal direction and vertical, both scaled up by a factor of four (peak horizontal acceleration equal to 0.68g). In this case, the analytical model accounted for the effects of vertical acceleration, variable normal load and velocity of sliding on the frictional properties of the bearings. It may be seen that the analytical results are in very good agreement with the experimental response.

COMPUTER CODES FOR THE ANALYSIS OF SLIDING ISOLATED STRUCTURES

Computer codes for the analysis of base-isolated building have been developed by Nagarajaiah, 1991 (program 3D-BASIS) and 1993 (program 3D-BASIS-TABS) and Tsopelas, 1991 (program 3D-BASIS-M) and 1994 (program 3D-BASIS-ME). These programs have the capability of modeling a variety of isolation elements including elastomeric and sliding bearings. The model of friction described in the previous section is incorporated in these programs. Program 3D-BASIS has been used in the analysis of a number of isolated structures, including the U.S. Court of Appeals in San Francisco (Amin 1993).

REFERENCES

1. American Association of State Highway and Transportation Officials-AASHTO (1991). "Guide specifications for seismic isolation design." Washington, D.C.

2. Amin, N., Mokha, A., and Fatehi, H. (1993). "Seismic isolation retrofit of the U.S. Court of Appeals building." Proc. ATC-17-1 Seminar on Seismic Isolation, Passive Energy Dissipation, and Active Control, March, San Francisco, CA.

3. Ciampi, V., and Marioni, A. (1991). "New types of energy dissipating devices for seismic protection of bridges." 3rd World Congress on Joint Sealing and Bearing Systems for Concrete Structures, Vol. 2, 1225-1245, Toronto.

4. Constantinou, M.C., Mokha, A.M. and Reinhorn, A. (1990). "Teflon bearings in base isolation. II: modeling," J. Struct. Engng., ASCE, 116(2), 455-474.

5. Constantinou, M.C., Kartoum, A., Reinhorn, A.M. and Bradford, P. (1991). "Experimental and theoretical study of a sliding isolation system for bridges." Report No. NCEER-91-0027, National Center for Earthquake Engineering Research, Buffalo, N.Y.

6. Constantinou, M.C., Tsopelas, P., Kim, Y-S and Okamoto, S. (1993). "NCEER-TAISEI corporation research program on sliding seismic isolation systems for bridges - experimental and analytical study of friction pendulum system (FPS)." Report No. NCEER-93-0020, National Center for Earthquake Engineering Research, Buffalo, N.Y.

7. Constantinou, M.C., Tsopelas, P. and Okamoto, S. (1994). "Experimental Study of a class of bridge sliding seismic isolation systems." 3rd U.S. - Japan Workshop on Earthquake Protective Systems for Bridges, January, Berkeley, CA.

8. Guéraud, R., Noel-Leroux, J. P., Livolant, M. and Michalopoulos, A.P. (1985). "Seismic isolation using sliding-elastomer bearing pads." Nuclear Engrg. and Design, 84, 363-377.

9. International Conference of Building Officials (1991). "Uniform building code, earthquake regulations for seismic-isolated structures". Whittier, CA.

10. Kawamura, S. Kitazawa, K., Hisano, M., and Nagashima, I. (1988). "Study on a sliding-type base isolation system--System composition and element properties." Proceedings of the 9th World Conference on Earthquake Engineering, pp. 735-740, Vol. 5, Tokyo and Kyoto, Japan.

11. Marioni, A. (1991). "Antiseismic devices on the Mortaiolo viaduct." 3rd World Congress on Joint Sealing and Bearing Systems for Concrete Structures. Vol. 2, 1263-1280, Toronto.

12. Martelli, A., Parducci, A. and Forni, M. (1993). "Innovative seismic design techniques in Italy." Proc. ATC-17-1 Seminar on Seismic Isolation, Passive Energy Dissipation, and Active Control, March, San Francisco, CA.

13. Medeot, R. (1991). "The evolution of aseismic devices for bridges in Italy." 3rd World Congress on Joint Sealing and Bearing Systems for Concrete Structures, Vol. 2, 1295-1320, Toronto.

14. Mokha, A., Constantinou, M.C. and Reinhorn, A.M. (1988). "Teflon bearings in aseismic base isolation: Experimental studies and mathematical modeling." Report No. NCEER-88-0038, National Center for Earthquake Engineering Research, Buffalo, NY.

15. Mokha, A., Constantinou, M.C., Reinhorn, A.M., and Zayas, V. (1991). "Experimental study of friction pendulum isolation system." J. Struct. Engng., ASCE, 117(4), 1201-1217.

16. Mokha, A., Constantinou, M.C. and Reinhorn, A.M. (1993). "Verification of friction model of teflon bearings under triaxial load." Journal of Struct. Engrg., ASCE, 119(1), 240-261.

17. Nagarajaiah, S., Reinhorn, A.M., and Constantinou, M.C. (1991). "3D-BASIS: Nonlinear dynamic analysis of three-dimensional base isolated structures." Report No. NCEER-91-0005, National Center for Earthquake Engineering Research, Buffalo, NY.

18. Nagarajaiah, S., Li, C., Reinhorn, A.M. and Constantinou, M.C. (1993). "3D-BASIS-TABS computer program for nonlinear dynamic analysis of three-dimensional base isolated structures." Report No. NCEER-93-0011, National Center for Earthquake Engineering Research, Buffalo, NY.

19. Pavot, B. and Polust, E. (1979). "Aseismic bearing pads." Tribology International, 12(3), 107-111.

20. Plichon, C., Guéraud, R., Richli, M.H. and Casagrande, J.F. (1980). "Protection of nuclear power plants against seism." Nuclear Technology, 49, 295-306.

21. Soong, T.T. and Constantinou, M.C. (1992). "Base isolation and active control technology-case studies in the U.S.A." IDNDR International Symbosium on Earthquake Disaster Reduction Technology, Building Research Institute, Ministry of Construction, Tsukuba, Japan, December.

22. Tsopelas, P.C., Nagarajaiah, S., Constantinou, M.C., and Reinhorn, A.M. (1991). "Nonlinear dynamic analysis of multiple building base isolated structures. Program 3D-BASIS-M." Report No. NCEER-91-0014, National Center for Earthquake Engineering Research, Buffalo, NY.

23. Tsopelas, P., Okamoto, S., Constantinou, M.C., Ozaki, D. and Fujii, S. (1994). "NCEER-TAISEI corporation research program on sliding seismic isolation systems for bridges - experimental and analytical study of systems consisting of sliding bearings, rubber restoring force devices and fluid dampers," Report No. NCEER-94-0002, National Center for Earthquake Engineering Research, Buffalo, N.Y.

24. Tsopelas, P., Constantinou, M.C. and Reinhorn, A.M. (1994)." Extended program 3D-BASIS-ME for nonlinear dynamic analysis of single and multiple building seismically isolated structures and seismically isolated liquid storage tanks." Report No. NCEER-94-xxxx, National Center for Earthquake Engineering Research, Buffalo, NY.

25. Zayas, V., Low, S.S. and Mahin, S.A. (1987). "The FPS earthquake resisting system, experimental report." Report No. UCB/EERC-87-01, University of California, Berkeley.

26. Zayas, V., Piepenbrock, T. and Al-Hussaini, T. (1993). "Summary of testing of the friction pendulum seismic isolation system." Proc. ATC-17-1 Seminar on Seismic Isolation, Passive Energy Dissipation, and Active Control, March, San Francisco, CA:

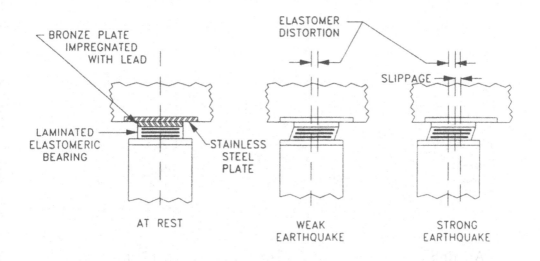

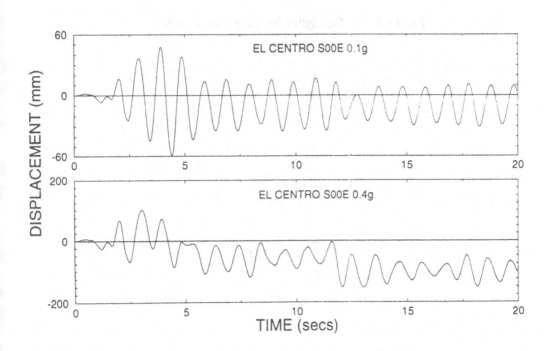

Figure 1   EDF System Configuration and Typical Response

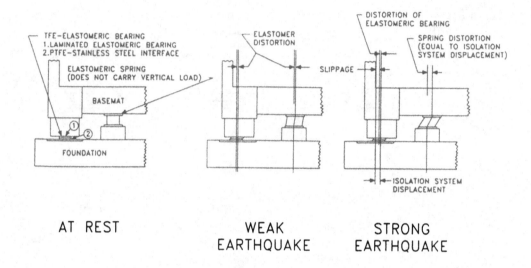

AT REST   WEAK   STRONG
                         EARTHQUAKE   EARTHQUAKE

Figure 2   Configuration of TASS System

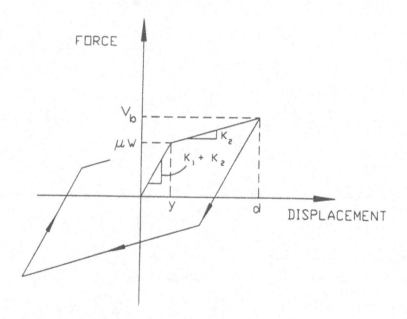

Figure 3   Idealized Force-displacement Relation of TASS System

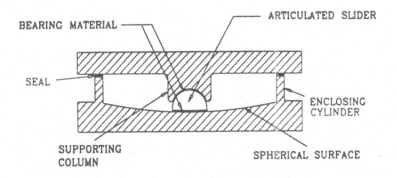

Figure 4   FPS Bearing Design

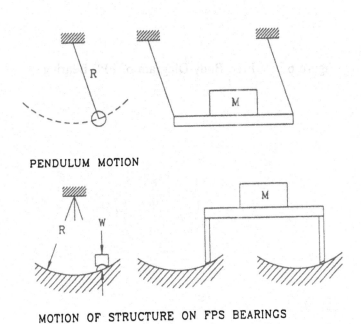

PENDULUM MOTION

MOTION OF STRUCTURE ON FPS BEARINGS

Figure 5   Basic Principle of Operation of FPS Bearing

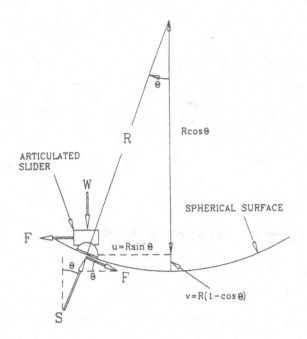

Figure 6   Free Body Diagram of FPS Bearing

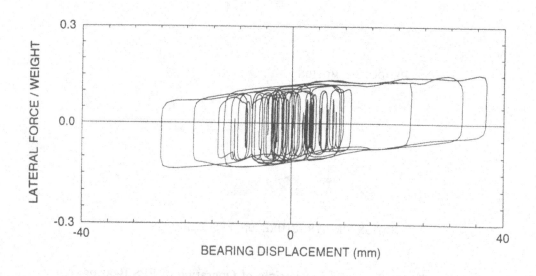

Figure 7   Recorded Loop of Force-displacement of FPS Bearing during Shake
Table Test. Excitation is 1940 El Centro Scaled to 0.68g

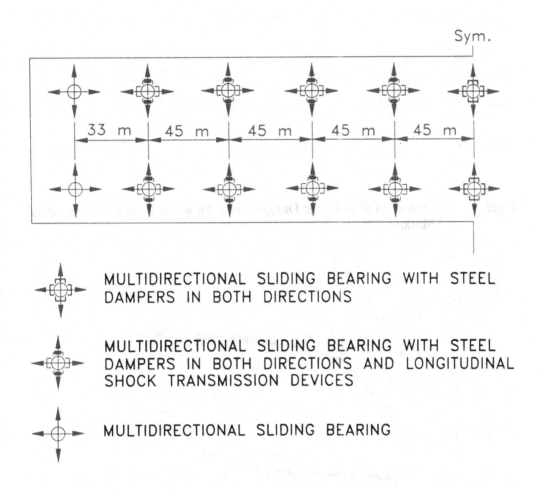

MULTIDIRECTIONAL SLIDING BEARING WITH STEEL
DAMPERS IN BOTH DIRECTIONS

MULTIDIRECTIONAL SLIDING BEARING WITH STEEL
DAMPERS IN BOTH DIRECTIONS AND LONGITUDINAL
SHOCK TRANSMISSION DEVICES

MULTIDIRECTIONAL SLIDING BEARING

Figure 8   Layout of Bearings in Mortaiolo Viaduct

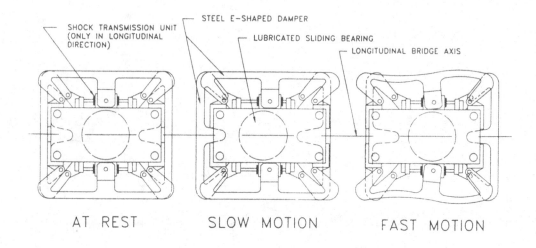

Figure 9   Operation of E-shaped Damper under Slow and Dynamic Longitudinal
            Motion

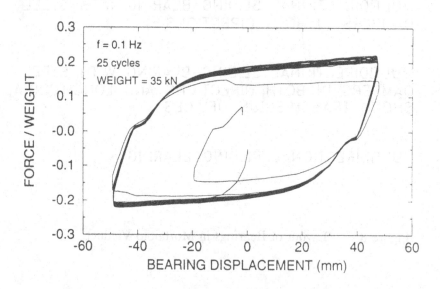

Figure 10   Recorded Force-displacement Relation of Scaled Lubricated PTFE
            Bearing with E-shaped Dampers

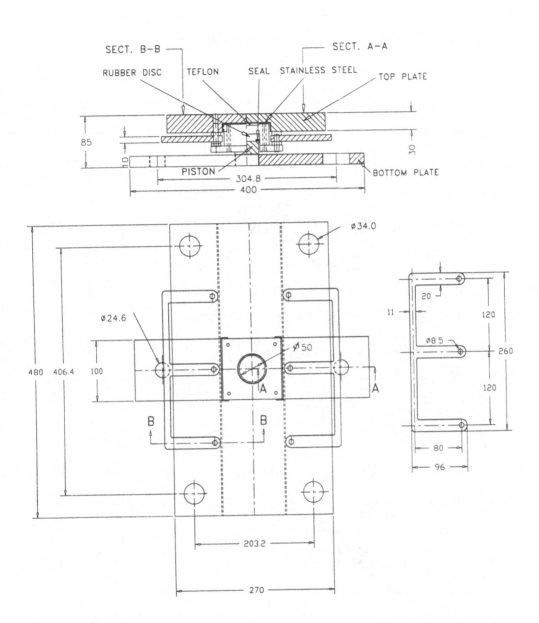

Figure 11   Design of Tested Bearing with E-shaped Dampers

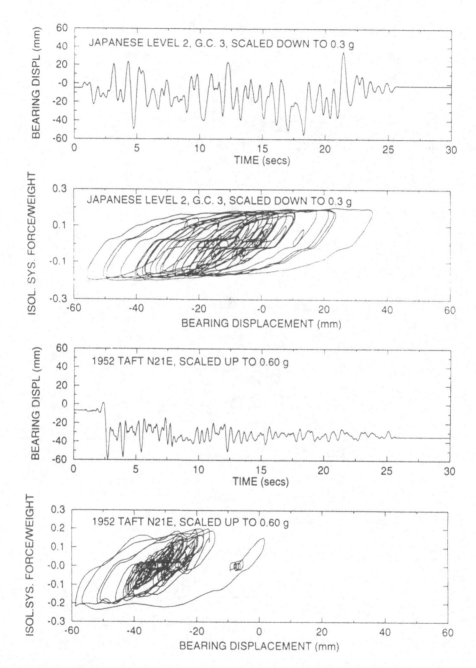

Figure 12   Recorded Response of Isolated Bridge Model with Lubricated
              Bearings and E-shaped Dampers to Japanese Level 2, G.C.3 and Taft
              Input

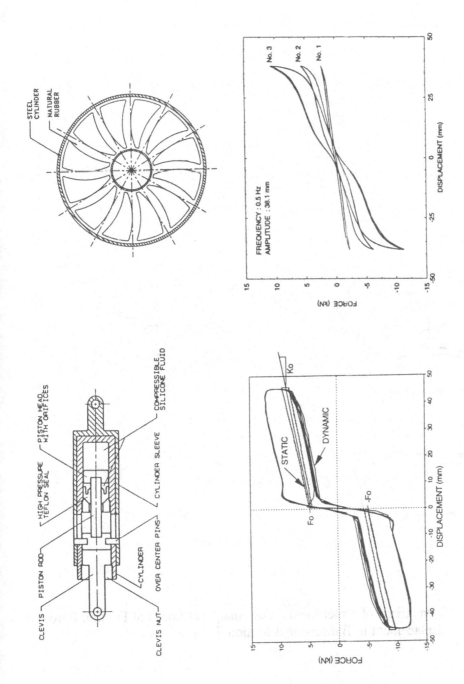

Figure 13   Unidirectional Compressible Fluid-Damping and Multidirectional Rubber Restoring Force Devices and Representive Force-Displacement Loops.

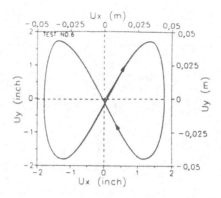

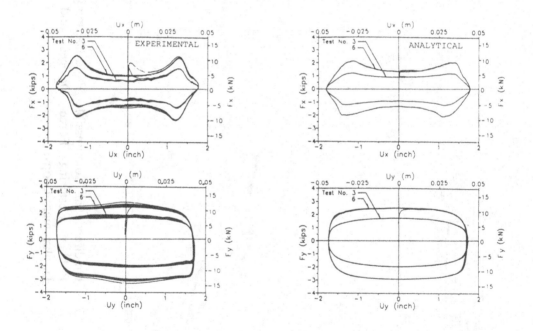

Figure 14   Comparison of Experimental and Analytical Loops of Friction Force-
displacement in Bidirectional Motion Tests

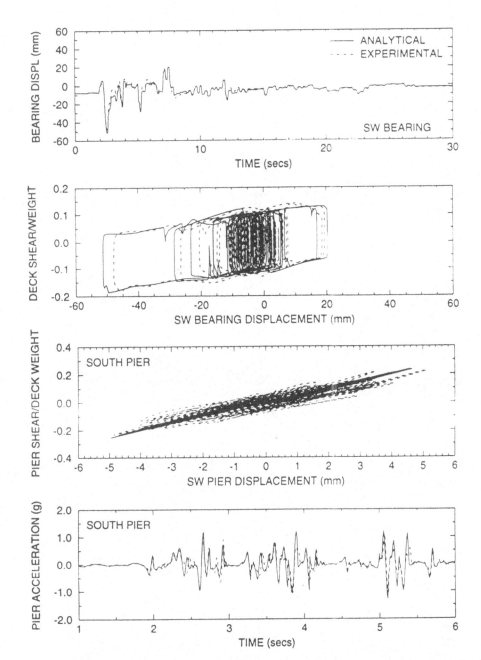

Figure 15   Comparison of Experimental and Analytical Response of Isolated
Bridge Model with FPS Bearings Subjected to 1952 Taft Motion Scaled
to 0.68g.

CHAPTER V

HYBRID EARTHQUAKE LOADING TESTS OF VARIOUS TYPES OF BASE ISOLATION BEARINGS

H. Iemura
Kyoto University, Kyoto, Japan

ABSTRACT

Under a joint research program between Kyoto University and the Hanshin Expressway Public Corporation, five isolation bearings of different damping mechanisms were tested under identical test conditions in order to evaluate their suitability and applicability to different types of bridges. The isolation bearings, including the lead-rubber bearing, two types of high-damping rubber bearings, sliding rubber bearing, and a conventional rubber bearing were tested under cyclic loading and hybrid earthquake (pseudo-dynamic) test methods. In addition to the fundamental cyclic load-deformation behavior, earthquake and resonant response behavior of the five isolation bearings were obtained. From the test results, equivalent damping ratios are evaluated based on the usual cyclic loading tests as well as from earthquake response of equivalent linear elastic models. In addition, the effectiveness of isolation is evaluated based on acceleration and displacement amplifications using earthquake response results obtained from the hybrid earthquake loading tests. In order to evaluate the effect of initial first-cycle stiffness of rubber bearings on the earthquake response, a second identical specimen of each type of isolation bearings was tested for its virgin earthquake response behavior.

INTRODUCTION

Research and development of seismic isolation bearings for bridges in Japan have dramatically intensified in the past two years. In Japan, several base-isolated buildings have already been designed and constructed. Many of these buildings were constructed by large construction companies and serve as company research centers or company dormitories. Acceptance of the seismic isolation concept to bridges will be easier since bridges are typically designed to be supported on bearings, whereas building structures are conventionally designed to be fixed at their base.

There will be more application of base isolation technology to bridges in the next few years with the completion of several fundamental and feasibility studies. An intensive test program between the Public Work Research Institute of the Japanese Ministry of Construction and 28 private companies [PWRI 1991; 1990] was conducted with the main objective of developing menshin devices for bridges and formulating design guidelines. The Hanshin Expressway Public Corporation also initiated research on applying seismic isolation to bridges and expressways under its management. Recently, menshin-type bearings have been installed to a steel box-girder bridge along the Bay Route in Osaka. A hybrid earthquake loading system had been developed at the Dept. of Civil Engineering, Kyoto University and was used to test a high-damping rubber isolation bearing for its fundamental cyclic behavior and earthquake response behavior [Iemura et al, 1991].

Under a joint research program between Kyoto University and the Hanshin Expressway Public Corporation, five isolation bearings of different isolation mechanisms were tested using this loading system in order to evaluate their suitability and applicability to different types of bridges. In this paper, the experimental results of the completed first phase of the research project will be presented.

VARIOUS TYPES OF BASE ISOLATION DEVICES

In buildings, base isolation devices are installed at the base to uncouple the whole building structure from destructive effects of earthquake ground motion. In bridges, the plane of isolation is below the girder superstructure so that inertial forces from the heavy superstructure that are transmitted to the critical pier supports can be reduced by the base isolation system (or more appropriately, seismic isolation system).

Many types of devices have been proposed to achieve seismic isolation. Several of these isolation systems use combinations of rubber bearings and complementary damping devices and have been used in buildings where these have the luxury of a whole basement story devoted to house several isolation devices spaced at considerable distances. However in bridges, where the bridge seats may not offer enough space, single-unit compact isolation devices are desirable. In addition, these devices must be made of robust construction due to exposure to harsher environmental conditions.

In recent years, many seismic isolation devices satisfying the above requirements for bridge application have been developed. Among these are the lead-rubber bearing, high-damping rubber bearing, and sliding rubber bearing. In a single unit are contained the necessary mechanisms to achieve seismic isolation: flexibility to lengthen the natural period of the supported structure, damping to limit excessive deformation, vertical stiffness to support heavy superstructure, and sufficient stiffness at low loads for serviceability against ambient vibrations. In the above-mentioned seismic isolation bearings, laminated elastomers made of rubber materials provide the required flexibility and interleaved steel plates give vertical stiffness. These bearings differ largely in the way damping mechanism is achieved. The lead-rubber bearing, which was developed in New Zealand, has a lead core at the center and utilize plastic yielding of the lead core for hysteretic damping. The high-damping rubber bearings use special rubber compounds that have inherent high damping. For the sliding rubber bearing, sliding friction provides the damping mechanism. Many other ways of providing damping have been proposed and developed, e.g., flexural yielding of steel bars, lead extrusion, torsional beam, high-yielding rubber core, etc. As base isolation concept gains more acceptance, new isolation techniques and configurations will continue to be developed. Excellent state-of-the-art reviews of the base isolation concept since its inception up to the present time can be found in Kelly [1986], Izumi [1988], Buckle and Mayes [1990].

For these isolation devices to be widely accepted by the structural engineering profession, their fundamental mechanical properties and their expected behavior during strong earthquakes should be well established. Extensive experimental tests are very much needed to study their behavior and provide data for analytical modeling and design.

The most fundamental test being done on seismic isolators is the repeated cyclic loading test (or quasi-static test). This is a test procedure to establish some mechanical properties of the isolators, such as hysteretic load-deformation behavior, cyclic energy absorption capacity, performance deterioration under repeated loads, etc.

The effectiveness of seismic isolators is very much dependent on the characteristics of the input earthquake motion and the supported structures. Most dynamic testing of seismic isolators have been done on shaking tables. However, the requirement of applying high vertical loads due to the high dependence of isolation characteristics on axial bearing pressure would necessitate a large-capacity table for a full-scale specimen; or specimens would have to be drastically scaled down in order to accommodate them in most shaking tables. Due to the limitation of shaking tables, one test was done using explosively simulated earthquake about 10 years ago [cited in Kelly, 1986].

During the last two decades, online hybrid test method (pseudo-dynamic test method) has become a very powerful tool to test for inelastic earthquake response of structures. Online hybrid test method is a computer-controlled experimental technique in which direct step-by-step time integration is used to solve the equations of motion. The computed displacement at each step is statically imposed on a specimen through computer-controlled load actuators in order to measure its restoring forces at the current deformation state. The measured restoring forces are then fed into the equations of motion to compute the next set of displacements. The numerical time integration process and the online loading process are carried out for a given earthquake ground motion history.

With the various types of isolation devices tested under different test methods and conditions, it is very difficult for the bridge designers to evaluate suitable isolation devices for a particular bridge. In addition, most of the tests conducted on large-scale specimens give only cyclic load-deformation characteristics and very few earthquake response tests (mostly shaking-table tests). It is necessary to evaluate and verify the seismic performance of seismic isolation devices under earthquake excitations and for different structural configurations.

BENCHMARK TEST PROGRAM FOR VARIOUS ISOLATION BEARINGS

In order to evaluate suitable isolation devices for different types of bridges of the Hanshin Expressway Public Corporation, a joint research has been initiated with the Earthquake Engineering Laboratory of the Department of Civil Engineering of Kyoto University to evaluate the fundamental mechanical properties and earthquake response behavior of various isolation devices of different mechanisms tested under similar conditions.

The isolation bearings tested were lead-rubber bearing (LRB), two types of high-damping rubber bearings (HDR1, HDR2), and sliding rubber bearing (SRB). In addition, a conventional rubber bearing (CRS) was also tested and used for comparison with the base isolation devices. Two identical specimens were made for each type of bearings. The details of the specimens are shown in Fig. 1.

The test set-up used in this experiment is shown in Fig. 2. The specimen is bolted to the underside of the rigid load-transfer beam and to a rigid platform that is fixed to the rails of the strong reaction floor. Different heights of isolator specimens could be accommodated by inserting filler plates between the specimen and the test platform.

The actuator for horizontal motion (labeled no. 1) has a maximum stroke of ± 125 mm and a maximum load capacity of 40 tonf. Each of the vertical actuators (no. 2 and no. 3) has load capacity of 40 tonf; hence, vertical load as high as 80 tonf can be applied on a specimen. Under the present set-up, specimens can be loaded to about $\pm 150\%$ shear strain. This will be extended to more than $\pm 250\%$ in the next phase of the experimental program with the procurement of a loading jack with larger stroke capacity.

The test methods adopted into the benchmark test program for this investigative study are mainly the cyclic loading tests and the hybrid earthquake loading tests. The notation for the test methods used for the discussion is shown in Table 1. For the earthquake response test, a 2-second SDOF system was assumed and subjected to the three earthquake records prescribed in the Specifications for Highway Bridges (S.H.B.) of Japan [1990]. The central difference time integration scheme is used for solving the equation of motion in this test. In the next phase of the research project, where substructured hybrid earthquake loading method will be used to test for earthquake response of seismically-isolated MDOF bridge system, a mixed explicit-implicit time integration scheme will be used for numerical stability and accuracy considerations.

Table 1   Test Methods and Notation

BCL	Bipolar Cyclic Loading Test	Cyclic Alternating Positive–Negative Loading
UCL	Unipolar Cyclic Loading Test	Cyclic Positive Loading
HE	Hybrid Earthquake Loading Test	Input Earthquakes from S.H.B.
HS	Hybrid Sweep Loading Test	Input Frequency-Sweeping Sine Base Acceleration

The sequence of tests is summarized in Table 2. Two specimens were fabricated for each type of isolation bearing. All tests can be done under one test set-up which enabled a series of test to be completed within a few hours and giving both fundamental mechanical properties and earthquake response behavior. To test for virgin earthquake response, the second specimen of each type was subjected to Type-3 Earthquake of S.H.B. This is especially important in that some rubber bearings exhibit exceptionally high stiffness during first displacement excursions. Due to this high initial stiffnesses, isolation effect might be diminished and consequently transmitting large forces to the pier supports.

Table 2   Sequence of Tests

	Test Sequence							
Reloaded	BCL-1	BCL-2	HE-1	HE-2	HE-3	HS	UCL-1	UCL-2
Virgin	HE-3 (×2)		BCL (×2)		HE-3			

TEST RESULTS AND DISCUSSIONS

Only a few representative results of the tests will be briefly described in this section. A more complete report on the test results is given in DPRC [1992].

Fundamental Cyclic Load-Deformation Behavior

First, results of first two sequences of cyclic loading tests for each type of isolation bearings are given in Figs. 3 to 7. Load-deformation behavior for SRB can be easily divided into that exhibited by the rubber bearing and that of the sliding surface as shown in Fig. 7. At the end of the first test series (Table 2), the rigid load-transfer beam is offset by 80 mm and fixed to enable large deformation in one orientation (UCL test). Figure 8 shows the results of the second UCL test for each type of isolation bearings. These serve as the tests for fundamental mechanical properties from which the usual equivalent shear moduli and equivalent damping ratios are computed. Fig. 9 gives the equivalent elastic shear moduli for each type of isolators, while Fig. 10 gives the equivalent damping ratios. The values G_{eq} and h_{eq} are both plotted against the shear strain ratio (total deformation divided by total height of rubber layers), except for the sliding rubber bearing in which the shear strain ratio is not defined. Instead, a normalized displacement parameter (total deformation divided by the design displacement specified) is used as shown in Fig. 10(b).

Earthquake Response Behavior

To evaluate earthquake response behavior, a 2-second SDOF system is subjected to the three earthquake records specified in S.H.B. (Fig. 11). Maximum ground acceleration amplitudes were scaled for each type of isolators to give shear strain ratios in the 100% – 150% range.

Representative results of earthquake response tests for each isolator type are given in Figs. 12–16. With these results, effectiveness of isolation can be evaluated directly in terms of earthquake response characteristics, or more specifically, the displacement and acceleration response time histories. From these, response reduction factors for both displacement and acceleration are computed and given in Table 3. In addition, linear elastic response is computed to match the maximum displacement and acceleration values from hybrid earthquake loading tests and obtain the equivalent damping ratios based on earthquake response behavior. Table 4 summarizes the equivalent damping ratios based on earthquake response tests.

For the second specimen of each isolation type, earthquake record no. 3 was used as input excitation (first test of second test series in Table 2) and virgin earthquake response for all isolation bearings were obtained. The main objective is to evaluate the effect of high initial stiiffnesses during first displacement excursions on earthquake response of seismically isolated bridge structures. This is specially pronounced in some types of rubber materials. Fig. 17 shows virgin earthquake response for one of the specimens. Comparison with the response of a preloaded specimen is indicated in Table 4.

Resonant Response Behavior

The same 2-second SDOF model was tested for resonant response when subjected to a sweeping sine-wave acceleration excitation (Fig. 18). Maximum input acceleration amplitudes were scaled to give resonant response also in the 100% – 150% range. The same technique as in hybrid earthquake loading test can be applied. Fig. 19 gives resonant response under sweeping waves for one of the specimens. Results for all of the types tested are summarized in Table 5.

CONCLUSIONS AND FURTHER CONSIDERATIONS

A benchmark test program for testing various seismic isolators with different damping mechanisms under similar conditions was adopted. Using cyclic loading test and hybrid earthquake loading test methods, it is possible to test for fundamental mechanical properties, earthquake response behavior and resonant response behavior. The effectiveness of isolation can be evaluated directly in terms of earthquake response values.

In the next phase of this research project, a loading jack with larger horizontal strokes ($\pm$250 mm) will be procured for the purpose of further examining the behavior at larger shear strain (above 200%). In addition, the seismic safety of the total structure will be investigated by substructured hybrid earthquake loading test to obtain the earthquake response of MDOF system. Using this technique, it will be possible to examine the effect of the isolation system on the response behavior of the other components of a bridge structure, especially the critical bridge piers.

ACKNOWLEDGEMENTS

The authors would like to acknowledge the following staff and students of the Earthquake Engineering Laboratory, Department of Civil Engineering, Kyoto University for their assistance and contribution to this research project: Mr. Shinji Nakanishi (laboratory engineer), Mr. Kazuyuki Izuno (research associate), Mr. Masaaki Tanaka (former graduate student), and Mr. Norio Watanabe (graduate student).

REFERENCES

I. Buckle, R. Mayes [1990]: "Seismic Isolation: History, Application, and Performance — A World View," *Earthquake Spectra*, Vol. 6, No. 2, 161–201.

Disaster Prevention Research Council (DPRC) [1992]: "Investigative Research on Seismic Isolation Design of Highway Bridges," Report submitted by the Earthquake Engineering Laboratory, Dept. of Civil Engg., Kyoto University to the Hanshin Expressway Public Corporation (in Japanese).

H. Iemura, Y. Yamada, W. Tanzo, Y. Uno, S. Nakamura [1991]: "On-line Earthquake Response Tests of High-Damping Rubber Bearings for Seismic Isolation," *1st US-Japan Workshop on Earthquake Protective Systems for Bridges*, Buffalo, New York, Sept.

M. Izumi [1988]: "State-of-the-Art Report on Base Isolation and Passive Seismic Response Control," *Proc., 9th WCEE*, Tokyo–Kyoto.

Japan Road Association [1990]: *Specifications for Highway Bridges*, Part V – Seismic Design (in Japanese).

J. Kelly [1986]: "Aseismic Base Isolation: Review and Bibliography, *Soil Dynamics and Earthquake Engineering*, Vol. 5, No. 3.

PWRI [1991, 1990]: "Report of Joint Research Between PWRI and 28 Private Firms on Development of Menshin Devices for Highway Bridges," *Technical Report of Cooperative Research*, No. 44 and No. 60, Public Work Research Institute.

W. Tanzo, Y. Yamada, H. Iemura [1992]: "Substructured Computer-Actuator Hybrid Loading Tests for Inelastic Earthquake Response of Structures," *Research Report 92-ST-01*, Kyoto University, January.

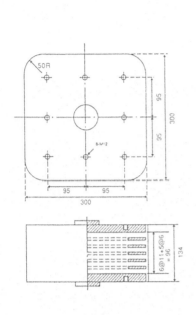

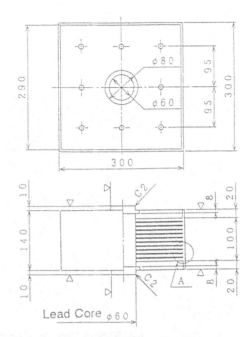

(a). Conventional Rubber Bearing   (b). Lead-Rubber Bearing

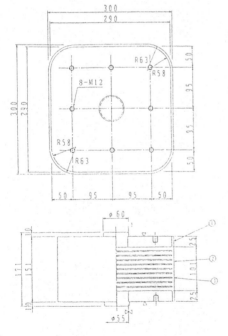

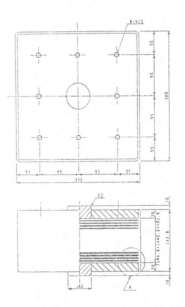

(c). High-Damping Rubber Bearing No. 1   (d). High-Damping Rubber Bearing No. 2

Figure 1   Tested Specimens (continuation next page)

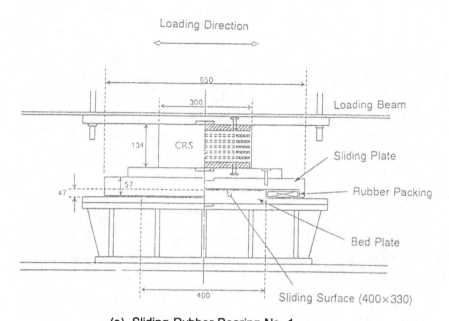

(e). Sliding Rubber Bearing No. 1

Figure 1   Tested Specimens (continued)

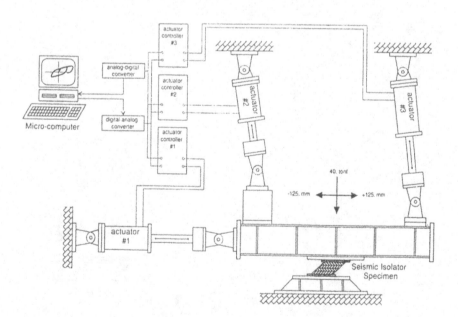

Figure 2   Test Set-up, Control and Measurement Systems

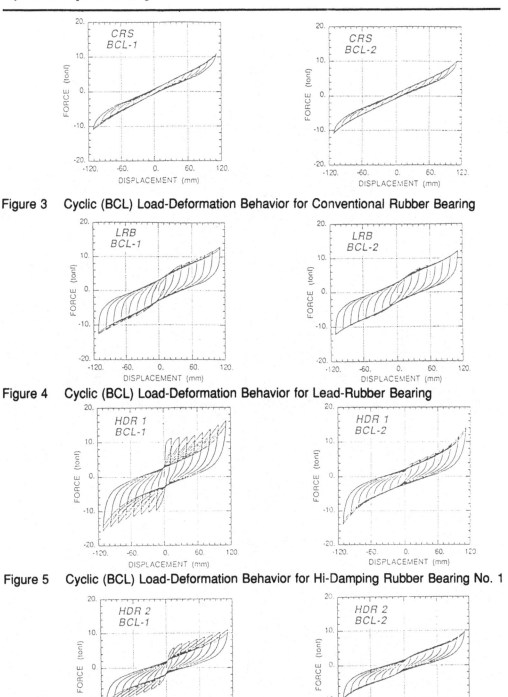

Figure 3   Cyclic (BCL) Load-Deformation Behavior for Conventional Rubber Bearing

Figure 4   Cyclic (BCL) Load-Deformation Behavior for Lead-Rubber Bearing

Figure 5   Cyclic (BCL) Load-Deformation Behavior for Hi-Damping Rubber Bearing No. 1

Figure 6   Cyclic (BCL) Load-Deformation Behavior for Hi-Damping Rubber Bearing No. 2

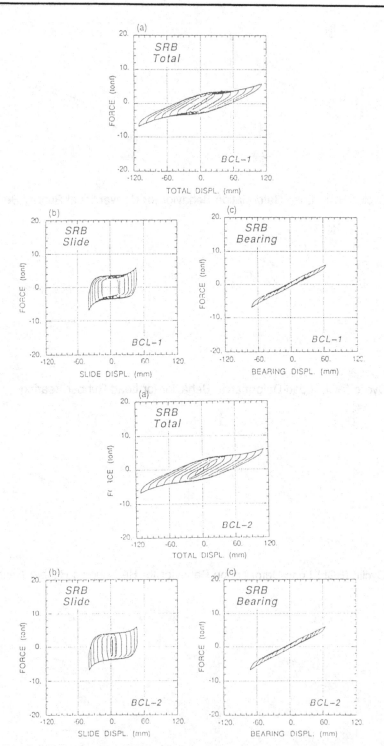

Figure 7   Cyclic (BCL) Load-Deformation Behavior for Sliding Rubber Bearing

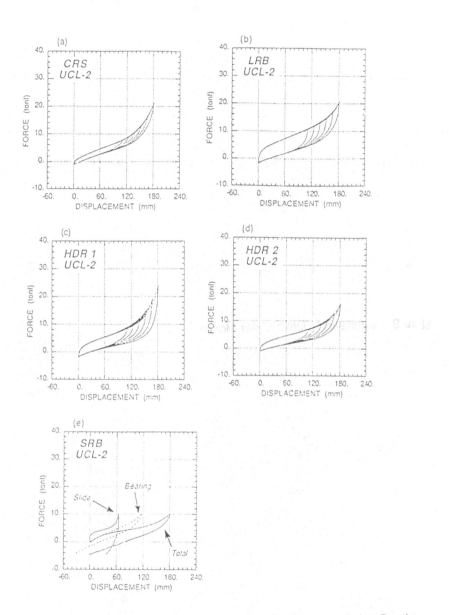

Figure 8   Cyclic (UCL) Load-Deformation Behavior of Various Isolation Bearings

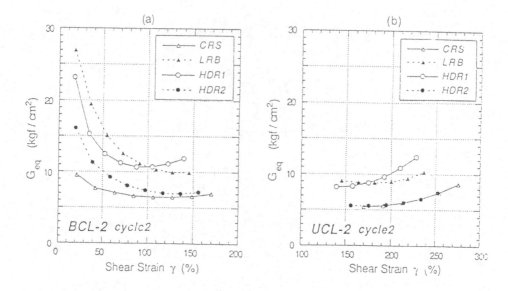

Figure 9   Equivalent Elastic Shear Moduli

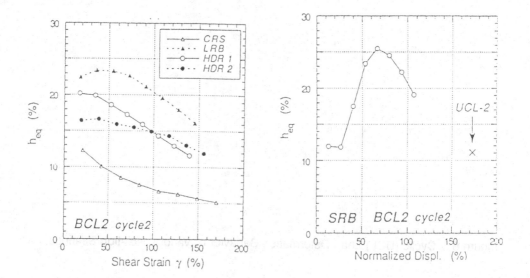

Figure 10   Equivalent Damping Ratios

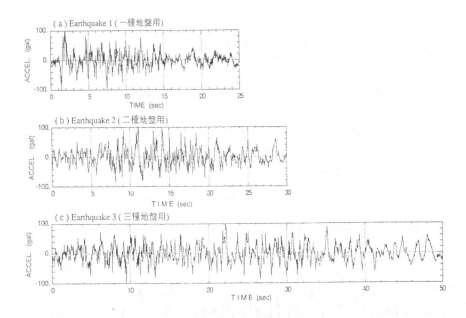

Figure 11   Earthquakes Records From the Specifications for Highway Bridges
and Used in the Hybrid Earthquake Loading Tests

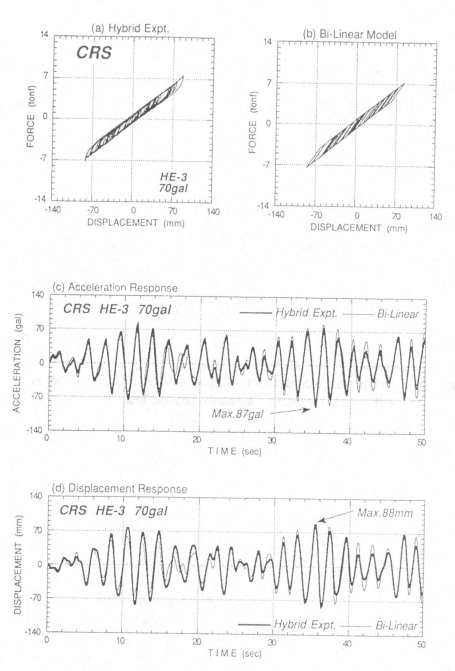

Figure 12   Earthquake Response from Hybrid Earthquake Loading Test (CRS, HE-3)

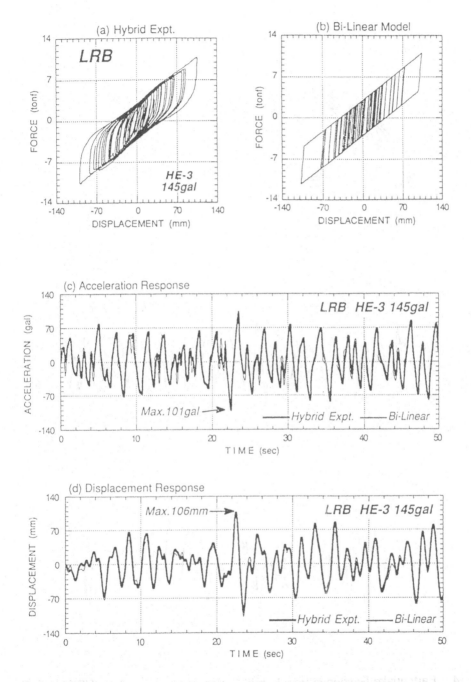

Figure 13   Earthquake Response from Hybrid Earthquake Loading Test (LRB, HE-3)

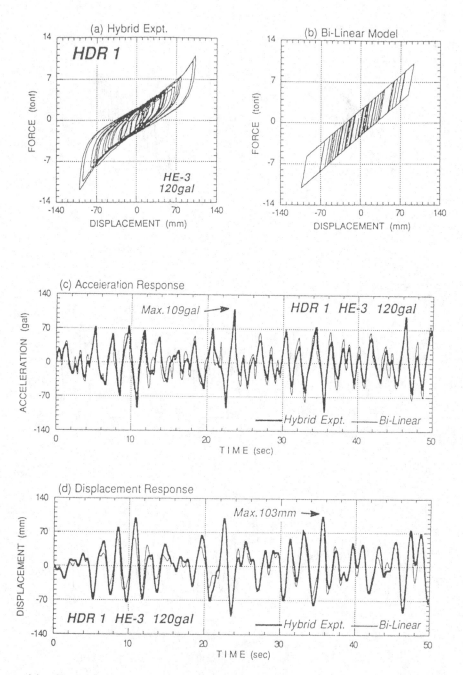

Figure 14   Earthquake Response from Hybrid Earthquake Loading Test (HDR1, HE-3)

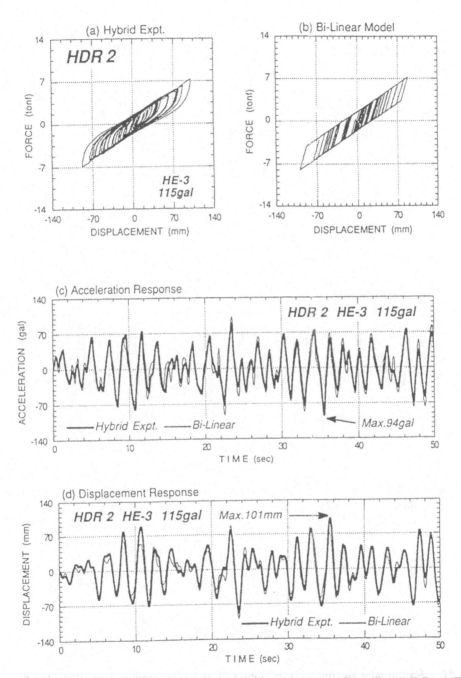

Figure 15   Earthquake Response from Hybrid Earthquake Loading Test (HDR2, HE-3)

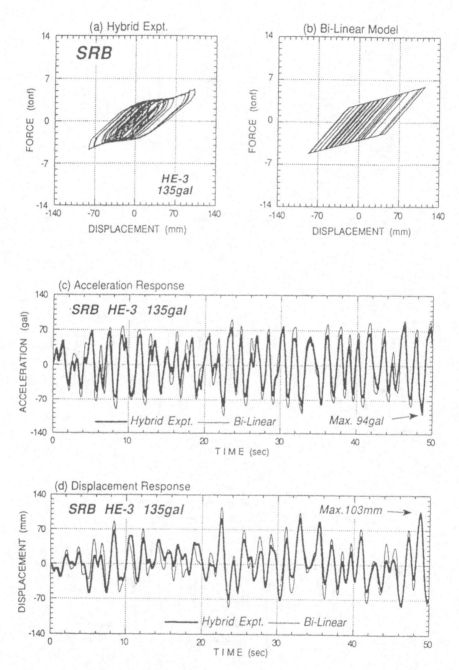

Figure 16   Earthquake Response from Hybrid Earthquake Loading Test (SRB, HE-3)

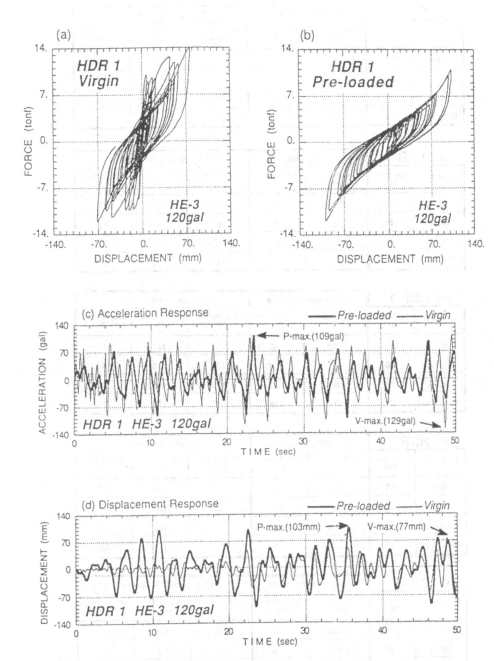

Figure 17   Virgin Earthquake Response from Hybrid Earthquake Loading Test

(a) CRS

Experiment	Input Accel. (gal)	Acceleration Response		Displacement Response		
		Max.Response (gal)	Amplification (%)	Max.Response (mm)	Shear Strain (%)	Magnification (%)
HE–1	100	101.1	101	97	147	96
HE–2	80	99.1	124	97	147	119
HE–3	70	86.8	124	87	132	123
HE–3 Virgin	70	88.9	127	83	126	117

(b) LRB

Experiment	Input Accel. (gal)	Acceleration Response		Displacement Response		
		Max.Response (gal)	Amplification (%)	Max.Response (mm)	Shear Strain (%)	Magnification (%)
HE–1	150	123.9	83	124	161	82
HE–2	140	98.9	71	103	134	73
HE–3	145	100.5	69	106	138	72
HE–3 Virgin	145	111.1	77	96	125	65

(c) HDR 1

Experiment	Input Accel. (gal)	Acceleration Response		Displacement Response		
		Max.Resp. (gal)	Amplification (%)	Max.Resp. (mm)	Shear Strain (%)	Magnification (%)
HE–1	120	119.8	100	113	141	93
HE–2	130	96.2	74	102	128	77
HE–3	120	109.4	91	103	129	85
HE–3 Virgin	120	129.2	108	77	96	63

(d) HDR 2

Experiment	Input Accel. (gal)	Acceleration Response		Displacement Response		
		Max.Resp. (gal)	Amplification (%)	Max.Resp. (mm)	Shear Strain (%)	Magnification (%)
HE–1	105	89.9	86	99	137	93
HE–2	105	86.2	82	93	129	87
HE–3	115	94.1	82	101	140	87
HE–3 Virgin	115	110.0	96	84	117	72

(e) SRB

Experiment	Input Accel. (gal)	Acceleration Response		Displacement Response		
		Max.Resp. (gal)	Amplification Factor (%)	Max.Resp. (mm)	Normalized Displ. (%)	Magnification (%)
HE–1	125	99.0	79	106	101	84
HE–2	130	102.6	79	105	100	80
HE–3	135	93.8	69	103	98	75
HS	48	108.9	227	113	108	232

Table 3   Response Reduction/Amplication Under Earthquake Loading

(a) CRS

Experiment	Maximum Response Matching		Substitute Damping
	Accel.match h_{eq} (%)	Displ.match h_{eq} (%)	h_s (%)
HE–1	5	6	9.0
HE–2	6	7	9.1
HE–3	7	7	9.1
HE–3 Virgin	7	8	–
HS	7	7	10.2

(b) LRB

Experiment	Maximum Response Matching		Substitute Damping
	Accel.match h_{eq} (%)	Displ.match h_{eq} (%)	h_s (%)
HE–1	14	12	30.5
HE–2	> 30	20	28.6
HE–3	> 30	18	28.2
HE–3 Virgin	18	21	–
HS	20	18	28.9

(c) HDR 1

Experiment	Maximum Response Matching		Substitute Damping
	Accel.match h_{eq} (%)	Displ.match h_{eq} (%)	h_s (%)
HE–1	5	6	21.8
HE–2	30	18	18.7
HE–3	12	13	18.3
HE–3 Virgin	9	22	–
HS	13	14	20.6

(d) HDR 2

Experiment	Maximum Response Matching		Substitute Damping
	Accel.match h_{eq} (%)	Displ.match h_{eq} (%)	h_s (%)
HE–1	11	6	22.1
HE–2	20	14	20.4
HE–3	15	12	18.3
HE–3 Virgin	11	17	–
HS	15	14	20.2

(e) SRB

Experiment	Maximum Response Matching		Substitute Damping
	Accel.match h_{eq} (%)	Displ.match h_{eq} (%)	h_s (%)
HE–1	19	11	25.2
HE–2	23	17	24.4
HE–3	29	16	25.8
HS	24	22	30.6

Table 4   Equivalent Damping Ratios Based on Earthquake Response

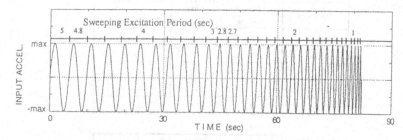

Figure 18   Input Frequency-Sweeping Sine Ground Acceleration Excitation

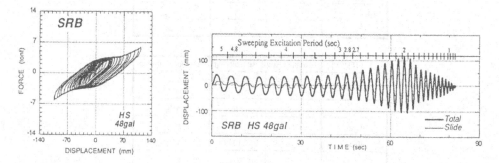

Figure 19   Resonant Response under Sweeping Excitation Input

HS – Experiment		Acceleration Response		Displacement Response		
Specimen	Input Accel. (gal)	Max.Response (gal)	Amplification (%)	Max.Response (mm)	Shear Strain (%)	Magnification (%)
CRS	15	87.0	580	88	133	580
LRB	37	98.6	266	105	136	280
HDR 1	28	103.5	370	100	125	352
HDR 2	30	102.9	343	104	144	342
SRB	48	108.9	227	113	★ 108	232

★ : Normalized Displ. (%)

Table 5   Response Reduction/Amplication Under Sweep Excitation

CHAPTER VI

EARTHQUAKE ENERGY PARTITIONING
IN BRIDGE STRUCTURES WITH SEISMIC ISOLATORS

H. Iemura

Kyoto University, Kyoto, Japan

Abstract

New earthquake energy spectra are proposed for practical design procedures using energy concepts. The spectra show integrated earthquake input energy of a unit mass for a given natural period and damping ratio. The total input energy and its partitioning in complex structures can be evaluated with the conventional procedure of modal analysis and the proposed spectra. Numerical simulations are carried out for inelastic structures to show the earthquake energy partitioning in time and in space. Inelastic earthquake response and earthquake energy partitioning of bridge structures with and without seismic isolators subjected to earthquake ground motions is evaluated to verify their engineering significance.

Earthquake Input Energy Spectra

The equation of motion of a single-degree-of-freedom inelastic structure with mass m, damping coefficient C, and hysteretic restoring force $F(x)$, subjected to earthquake ground acceleration $\ddot{z}$ is

$$m\ddot{x} + C\dot{x} + F(x) = -m\ddot{z} \qquad (1)$$

Multiplying by $dx(=\dot{x}dt)$ and integrating each term of Eq. (1) for the duration of earthquake ground motion $(0 \sim t_0)$, it follows that

$$\frac{1}{2}m\dot{x}^2(t_0) + \int_0^{t_0} C\dot{x}^2\,dt + \int_0^{t_0} F(x)\dot{x}\,dt = -\int_0^{t_0} m\dot{x}(t)\ddot{z}\,dt \qquad (2)$$

The first term on the left-hand side of Eq. (2) represents the kinetic energy of the mass at time t_0, the second term is the energy absorbed by viscous damping during vibration, and the third term includes both accumulated and dissipated energy by the hysteresis loops of the restoring force up to time t_0. On the right-hand side of Eq. (2) is the earthquake input energy to a structure from time 0 to t_0.

Partitioning of the earthquake energy in the time domain, represented by Eq. (2), is shown in Figs. 1 and 2 for a single DOF linear structure ($T = 0.5s, h = 0.03$) and a base-isolated structure ($T = 2.0s, h = 0.01$), respectively. The NS component of the El Centro record during the 1940 Imperial Valley Earthquake is used as input and hysteresis loops of a high-damping rubber bearing are used in the computer-actuator on-line hybrid earthquake response simulation (Iemura et al). Comparing the results, it is clearly found that the earthquake input energy E, the kinetic energy W_k and the potential energy W_E in the isolated structureare suppressed to a much lower level than in the non-isolated structure.

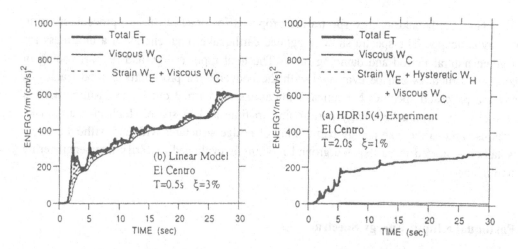

Fig. 1   Energy partitioning   Fig. 2   Energy partitioning
in a linear SDOF structure   in a base-isolated structure

The earthquake input energy to a linear SDOF system with a unit mass at the end of excitation is a function of the natural period and damping ratio. Analogous to conventional response spectra, it is named as "the earthquake input response spectra".

Energy Partitioning in Complex Structures

For a MDOF structure, its energy partitioning can be written as

$$\sum_i \left(\frac{1}{2} m_i \dot{x}_i^2\right) + \sum_i \int_0^{t_0} C_i \dot{y}_i^2 \, dt + \sum_i \int_0^{t_0} F_i \dot{y}_i \, dt = \sum_i \int_0^{t_0} (-m_i \dot{x}_i \ddot{z}) \, dt \qquad (3)$$

where i denotes the i-th mass or the i-th interstory, x and y denote relative displacement with respect to ground and interstory displacement, respectively. Eq. (3) can also be written as

$$\sum_i W_{Ki} + \sum_i W_{Ci} + \sum_i W_{Ei} + \sum_i W_{Hi} = \sum_i E_i \qquad (4)$$

At the end of response, kinetic and potential energies vanish, and thus the total input energy is absorbed by both viscous and hysteretic damping.

$$\sum_i W_{Ci} + \sum_i W_{Hi} = \sum_i E_i \qquad (5)$$

Using the modal analysis procedure and the proposed earthquake input energy spectra, the total earthquake input energy to an equivalent linear MDOF structure can be described by

$$\sum_i \int_0^{t_0} (-m_i \dot{x}_i \ddot{z}) \, dt = \sum_i \left\{ -m_i \int_0^{t_0} \left(\sum_s \phi_{is} \dot{y}_s\right) \ddot{z} \right\} dt = \sum_i m_i \left(\sum_s \phi_{is} E_s\right) \qquad (6)$$

where E_s is the earthquake input energy in s-th mode. E_s can be found from the proposed energy spectra. When there is no modal coupling, the earthquake input energy estimated by Eq. (5) agrees perfectly with the results of step-by-step numerical calculation. Hence, analytical estimation of earthquake input energy and its partitioning can be carried out with application of equivalent linearization techniques.

Inelastic Earthquake Response of Different Bridge Models

For numerical simulation of earthquake energy partitioning in bridge structures with and without seismic isolators, one of typical bridges used by Hanshin Expressway Public Corporation shown in Fig. 3(a) is adopted. A bridge consists of steel girder, reinforced concrete (RC) pier and pile foundation. They are modeled with flexure beam elements as shown in Fig. 3(b). Trilinear hysteretic restoring force characteristic is assumed for moment curvature $(M - \phi)$ relation of a pier. Pile foundation is modeled by horizontal and rotational elastic springs in accordance with the earthquake design specification of highway bridges in Japan. Energy dissipation due to soil-structure interaction is modeled by 10% viscous damping.

The following four types of connection between the girder and the pier top are modeled and inserted between No. 12 and 10 mass in the model shown in Fig. 3(b).

1. Rotation-free pin connection (Model 1)
2. Elastic spring support for which the natural period is set at 2.0 sec (Model 2)
3. Isolator with bilinear hysteretic restoring force (Model 3)
4. Elastic spring support with viscous damper (Model 4)

As earthquake ground input motion, NS-component of th El Centro record during the 1940 Imperial Earthquake is adopted and maximum acceleration is scaled to 600 gal to represent an extremely large event.

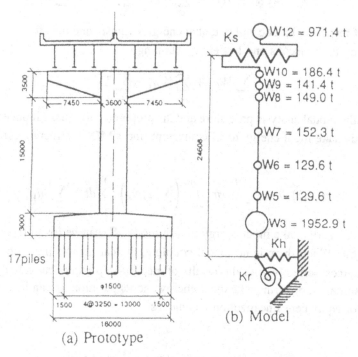

(a) Prototype   (b) Model

Fig. 3   Prototype and model of a bridge structure

Displacement response of a girder which is pin connected to a pier (Model 1) and moment-curvature $(M - \phi)$ hysteretic response of section No. 4 of pier bottom are shown in Fig. 4(a) and (b), respectively. Maximum displacement response of a girder is found to be 15.3 cm with vibration period of one second. Bending moment response of a section at the pier bottom goes deeply into plastic region, and deterioration of both resistance and stiffness is observed. Displacement and $(M - \phi)$ response of a bridge with elastic spring support (Model 2) are shown in Fig. 5(a) and (b). Although large displacement (35 cm) is found due to soft elastic spring support, the moment response at the pier bottom shows much less hysteretic behavior. Fig. 6(a) and (b) show displacement and $(M - \phi)$ response of a bridge with an isolator with bilinear hysteretic restoring characteristics (Model 3). Fig. 7(a) and (b) show response of a bridge with elastic spring support and viscous damper (Model 4). In both models 3 and 4, much smaller hysteretic behavior is found compared to pin connection. Maximum displacement (20 cm for Model 3, 10 cm for Model 4) is also found smaller than Model 2.

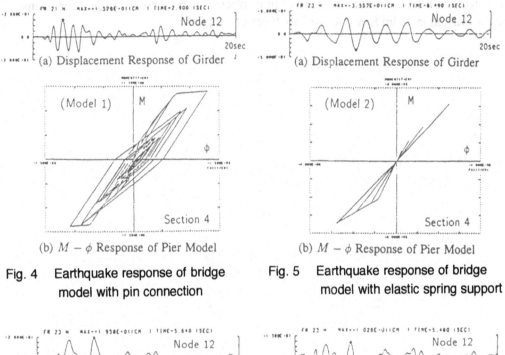

(a) Displacement Response of Girder

(b) $M - \phi$ Response of Pier Model

Fig. 4   Earthquake response of bridge model with pin connection

(a) Displacement Response of Girder

(b) $M - \phi$ Response of Pier Model

Fig. 5   Earthquake response of bridge model with elastic spring support

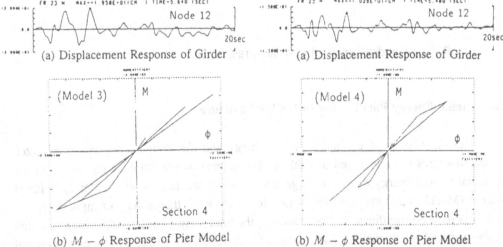

(a) Displacement Response of Girder

(b) $M - \phi$ Response of Pier Model

Fig. 6   Earthquake response of bridge model with bilinear hysteretic isolator

(a) Displacement Response of Girder

(b) $M - \phi$ Response of Pier Model

Fig. 7   Earthquake response of bridge model with viscous damper

The spatial distribution of maximum displacement response and acceleration and maximum bending moment response is plotted Fig. 8(a), (b), and (c). The displacement and acceleration response of a girder and bending moment response of a pier with pin connection (Model 1) show much larger response than others (Models 2, 3 and 4). This verifies that pin connection gives higher seismic load to piers than other models. It should be noted that a higher variation is found in displacement response of girders for seismic safety evaluation of bridges with various types of supports.

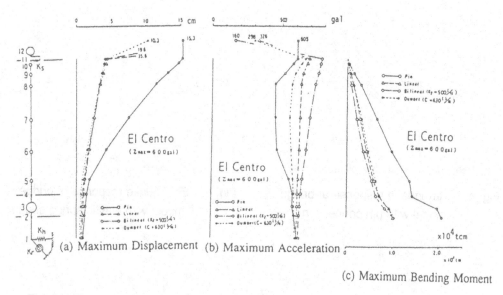

Fig. 8   Spatial distribution of maximum response

Earthquake Energy Partitioning in Bridge Structure

The partitioning of earthquake input energy and absorbed energy in the analyzed bridge structures is summarized in Table 1. The superstructure shares large portion of the earthquake input energy due to its large mass. When elastic support (Model 2), bilinear isolator (Model 3) and viscous damper (Model 4) are used, the earthquake input energy to girders is reduced significantly. Consequently, the total input energy to the bridge structure is also reduced significantly. The reason of this reduction is due to longer vibrational period of softly-supported superstructures.

Earthquake input energy is absorbed by different structural members depending on different structural models. In pin connection (Model 1), earthquake response of a pier goes into inelastic region deeply and absorbs one-fifth of the total input energy, which means that the pier has to accept some earthquake damage. In Model 2 of elastic support, all elements stays in elastic range and almost all of input energy is absorbed by soil-structure interaction. In Models 3 and 4, from one-third to half of input energy is absorbed by the isolator or the damper. Consequently, hysteretic energy absorption by piers is reduced significantly which would then guarantee high reliability of bridge structures.

In all models, large portion of the input energy is absorbed by soil-structure interaction. The simulated results of this study, including different level and types of earthquake response, will be used for practical design of energy-absorbing capacity of seismic isolators.

Table 1   Spatial partitioning of earthquake input energy and absorbed energy in bridge structures

Energy Partioning	Types of Supports	Pin Connection	Elastic Supports	Bilinear Isolator	Viscous Damper
Earthquake Imput Energy	Super Structures	22 550	10 400	9 912	16 460
	Piers	14 070	8 473	10 130	10 550
	Foundations	9 613	6 556	7 947	6 752
	Total	46 230	25 430	27 990	33 760
Hysteretic Energy Absorption	Bearings	0	0	8 661	16 710
	Piers	10 210	229	46	293
	Foundations	0	0	0	0
	Sub Total	10 210	230	8 710	17 000
Viscous Energy Absorption		36 860	25 200	19 180	16 750
Total of Absorbed Energy		47 070	25 430	27 890	33 750

Unit : (t · cm)

Conclusions

To develop earthquake energy based design concepts, new earthquake energy spectra are proposed. Analytical control of partitioning of the earthquake energy in base-isolated structures has become possible with the use of the spectral and modal analysis of equivalently linearized systems. Effects of inelastic behavior of isolators and of dynamic dampers are evaluated from numerical simulations. Appropriately designed isolators are found not only to suppress earthquake input energy but also to absorb most of it, thus, reducing the structural damage significantly.

Acknowledgements

The author would like to thank Dr. Kazuyuki Izuno (Assoc. Prof., Ritsumeikan Univ.) and Dr. William Tanzo (Research Assoc, Saitama Univ.) for their help in preparation of this lecture note.

References

Iemura, H. (1984). "Hybrid Experiments on Earthquake Failure Criteria of Reinforced Concrete Structures," *Proc. 6th WCEE*, VI, 103–110.

Iemura, H., et al (1982). "Reduced DOF Approximations for Earthquake Damage Assessments of Deteriorating Hysteretic Structures," *Proc. 6th JEES*, 1233–1240, Dec.

H. Iemura, H., et al (1991) "On-line Earthquake Response Tests of High-Damping Rubber Bearings for Seismic Isolation," *Proc. 1st US-Japan Workshop on Earthquake Protective Systems for Bridges*, Buffalo NY, Sept.

Kato, B. and Akiyama, H. (1985). "Energy Input and Damages in Structures Subjected to Severe Earthquakes," *Proc., AIJ*, No. 235, pp.9–18. (in Japanese)

Ohno, T., et al. (1983). "Quantitative Estimation of Plastic Energy Absorbed in Structures Subjected to Seismic Excitation," *Proc., JSCE*, No. 333, 91–99, May. (in Japanese)

CHAPTER VII

SEISMIC ISOLATION DEVELOPMENT IN U.S.: CASE STUDIES

M.C. Constantinou
State University of New York at Buffalo, Buffalo, NY, USA

ABSTRACT

This section describes the development and application of seismic (or base) isolation in the United States. Significant strides have been made in the implementation of the technology. This is attributed to recent advances in seismic isolation hardware and the removal of a number of impediments, which delayed widespread implementation in the near past.

INTRODUCTION

Recent advances in isolation system hardware and the potential advantages offered by seismic isolation resulted in the construction of a number of isolated structures in the U.S. In the past, a number of impediments delayed the implementation of this technology until 1986, when the first isolated building, the

Foothill Communities Law and Justice Center in San Bernardino, CA, was constructed.

These impediments (Mayes 1990) have been now largely removed. In particular,

(a) The lack of code provisions for isolated structures has been resolved by the approval and incorporation in the Uniform Building Code (ICBO 1991) of "Earthquake Regulations for Seismic-Isolated Structure." Similar guidelines have been developed for bridges (AASHTO 1991).

(b) The initial construction costs do not necessarily dominate the decision making process. Rather, considerations such as functionality during and after a strong earthquake, improved building performance, increased safety to occupants and historical and architectural value of existing buildings play a significant role in the decision making process.

(c) Comparable isolation systems are now available that allow competitive bidding on Federal Government funded projects. For example, in recent Federal Government projects, high damping rubber bearings, lead-rubber bearings and sliding friction pendulum system (FPS) bearings participated in competitive bidding.

(d) The Federal Government and the California State Government provide leadership in the implementation of seismic isolation. To date, the Federal Government, through the General Services Administration, has funded one major retrofit isolation project, the U.S. Court of Appeals in San Francisco, and is expected to fund other retrofit projects. Furthermore, the California State Government has approved the isolation retrofit of the State of California Justice Building in San Francisco. These structures are of significant historical and architectural value.

DEVELOPMENT OF SEISMIC ISOLATION SYSTEMS IN THE U.S.

Research on the development of seismic isolation systems for buildings began at the University of California at Berkeley in 1976 (Kelly 1991). Shrotly after, the California Department of Water Resources installed elastomeric isolators to a number of 230 kV circuit breakers in Southern California. The isolators were low damping natural rubber bearings of the French Gapec type (Kircher 1979).

The research program at U.C. Berkeley received funding from the Malaysian Rubber Producers Research Association (MRPRA) in England, the U.S. National Science Foundation and the Electric Power Research Institute and concentrated on the development of elastomeric isolation systems for the protection of buildings and secondary systems attached to them. Low damping elastomeric systems were originally studied. Subsequently, energy dissipating elements (such as lead plugs in the bearings, yielding steel devices, frictional systems) were combined with elastomeric bearings and experimentally studied (Kelly 1980a, 1980b, 1981, 1982, 1990). In 1985, we have the first application of lead-rubber bearings in the retrofit of the Sierra Point overhead, a bridge near San Francisco.

The researchers at U.C. Berkeley determined that a desirable method of increasing damping in the isolation system is to provide it in the rubber itself. A high damping rubber compound was developed by MRPRA and used in the first seismically isolated building in the United States. The Foothill Communities Law and Justice Center in San Bernardino was completed in 1986.

At just about the time of completion of the Foothill Communities building, research at U.C. Berkeley was conducted on a novel sliding isolation system (Zayas 1987). In 1987 experimental research began at the University at Buffalo on sliding isolation systems. With funding from the National Science Foundation, the National Center for Earthquake Engineering Research (NCEER) and industry, several sliding isolation systems were studied (Mokha 1988 and 1991, Constantinou 1991a and 1991b). Furthermore, the researchers at U. Buffalo developed the first computer programs specifically designed for the analysis of seismically isolated structures (Nagarajaiah 1991, Tsopelas 1991). Moreover, the researchers at U. Buffalo studied combined vibration and seismic isolation systems (Makris 1992) and equipment isolation systems (Demetriades 1992).

In 1988, a sliding isolation system, the FPS system (Zayas 1987), found its first application in the United States. A 50-year old, 30 m tall water tank near San Francisco was fitted at its base with FPS bearings which were designed with uplift restraint.

In the 1990s joint research efforts between U.S. institutions and the Japanese construction industry began. The Earthquake Engineering Research Center of U.C. Berkeley and Shimizu Corporation, with the participation of MRPRA and Angonne National Laboratory, conducted evaluation studies of elastomeric isolation systems (Kelly 1993). NCEER and Taisei Corporation began work on the development and evaluation of advanced sliding isolation systems (Constantinou 1992a, 1992b, 1993a, 1993b, 1994; Tsopelas 1994). This project involved the participation of U.S. industries which manufacture isolation hardware for aerospace and military applications. Fluid damping and restoring force/damping devices have been adapted for use in seismic isolation systems, tested and found to be very efficient. Very recently, fluid damping devices have been specified for an isolated structure in California.

SEISMICALLY ISOLATED STRUCTURES IN THE U.S.

A pre-1989 review of seismic isolation activity in the U.S. and worldwide has been presented by Buckle 1990. The occurrence of the 1989 Loma Prieta earthquake in San Francisco appears to have accelerated the implementation of seismic isolation to buildings. In particular, a number of seismic-isolation retrofit projects in the San Francisco area have been completed, are on-going, or underwent the conceptual design phase and await funding. In bridges we have accelerated implementation of seismic isolation systems as of 1991. Prior to 1991, only seven bridges were isolated of which five were in California. With the adoption of seismic design specifications by AASHTO, which apply for all parts of the country, and the development of seismic isolation guidelines for bridges (AASHTO 1991), the implementation accelerated so that by the end of 1992 there were 33 isolated bridges. Interestingly, the majority of these bridges (26) were in States other than California.

To date, there are in the U.S. 25 seismically-isolated buildings (constructed or under construction), 59 isolated bridges, a number of isolated equipment and two isolated tanks. Several more are in the design or planning stage. Table 1 presents a directory of seismically isolated buildings in the U.S. as of early 1994. The table includes only buildings which have been constructed or are in the construction phase. Studies have been conducted on a number of other buildings and decisions have been made to proceed with seismic isolation when funding becomes available. Tables 2 and 3 list isolated bridges and other structures which were constructed or were in the construction stages as of late 1993.

The directories in Tables 1 to 3 illustrate the range of seismic isolation products currently available in the U.S. These include high damping rubber bearings, lead-rubber bearings, friction pendulum system (FPS) bearings, PTFE sliding bearings and fluid viscous dampers. Furthermore, spring-viscous damper systems have been used for combined seismic and vibration isolation (Makris 1992).

SEISMIC ISOLATION RETROFIT PROJECTS

Retrofitting buildings by seismic isolation appears to be a unique U.S. practice. Table 4 presents a summary of isolation retrofit projects. The table includes information on the structural system, isolation system and the dominant factors, other than that of better performance, which motivated owners to select isolation. These motivating factors include:

(a) *Historical Building Preservation.* Modification or demolition of historical building features is minimized.

(b) *Functionality*. The facility remains operational during and after an earthquake.

(c) *Design Economy*. Seismic isolation is the most economical retrofit measure.

(d) *Investment Protection*. Long-term economic loss is reduced.

(e) *Content Protection*. Important contents (e.g., artifacts) are protected against damage.

The retrofitted isolated structures range from small buildings to monumental ones with floor areas up to 32,000 m² and building weights up to 55,000 metric tons. Furthermore, two structures incorporate a unique feature in their isolation systems. The Hawley Apartments building has its isolators placed at the bottom of each column without interconnection of the columns to form an isolation basemat. This reduced construction costs and did not affect the ceiling to floor height. The Rockwell building incorporates a similar feature with the isolators placed close to the bottom of the second story columns.

The design criteria employed for each of the retrofit projects depended on the type of structure and desired performance. In general, these criteria conformed with and in many cases exceeded those of the Uniform Building Code (ICBO 1991). This was necessitated either by the severity of operational objectives or by special damage protection goals.

Further details are presented in the sequel for two case studies. Both involve monumental structures of historical and architectural value.

CASE STUDIES IN SEISMIC ISOLATION

Salt Lake City and County Building, Salt Lake City, Utah

This represents the first building to be retrofitted using seismic isolation. The structure was built in 1894, has plan dimensions of 80m by 40 m and extends over five stories to a height of 30 m. A clocktower extends another 37 m above the roof of the main building. It was constructed of unreinforced brick masonry and sandstone. The structure is monumental and highly ornamented with a total weight of about 35,250 metric tons.

At the time the building was constructed earthquake engineering was nonexistent. Over the years, occasional earthquakes caused cracks, loosened stonework and shifted decorative statues atop of the clocktower. The condition of the building became an official concern and in early 1970 studies of the building condition were conducted. The studies recommended stiffening of the clocktower using a steel space truss. Continued deterioration of the building prompted a new comprehensive study in early 1980. The study recommended saving the building because of its historical and architectural significance. Various approaches for seismic retrofit were considered.

All approaches to retrofit, except that of seismic isolation, were destructive to the architecture of the building and capable of only preventing building collapse. Seismic isolation was finally selected despite the fact that it was not the least expensive alternative. At that time, engineers in the U.S. had confidence only in the use of elastomeric isolation systems. Accordingly, the Request for Proposals in 1986 called for competitive bids on elastomeric isolation systems. The successful bidder proposed the use of a combination of 447 lead-rubber and plain rubber bearings (Mayes 1988).

The building loads were transferred and distributed to the isolators by an elaborate system, consisting of:

-Concrete side beams cast on either side of the masonry walls above the isolators, post-tensioned and clamped through the walls.

-Concrete cross beams cast on top of the isolators.

-Steel grillage beneath the isolators to transfer load to the footings.

-A concrete floor above the isolators to create rigid diaphragm action.

U.S. Court of Appeals, San Francisco, California

This building was constructed in two phases. Construction of the original building was completed in 1905. An addition was completed in 1935. The building has plan dimensions of 95 m by 81 m and overall height of 25 m above street level. It consists of a full basement, four floors and two mezzanines. It weight is approximately 55,000 metric tons.

The superstructure of both the 1905 and 1933 buildings is steel framed with the columns consisting of laced pairs of channels. The floor framing consists of cinder reinforced concrete arches between floor beams which are riveted to the girders. The foundation consists of individual footings in the 1905 building and of 10 m-long concrete piles in the 1933 building. The exterior of the building features granite walls which are fastened to the steel framing above the second floor level. The first story exterior granite walls are supported on brick masonry basement walls.

The building represents an early example of American Renaissance Style without peer in San Francisco and the Western U.S. It has both significant historic and architectural merit.

The U.S. Court of Appeals building suffered damage to its interior tile walls and exterior granite walls during the 1989 Loma Prieta earthquake. The interior walls formed a part of the lateral load resisting system and their damage significantly reduced the seismic resistance of the building. The building was evacuated and a study was undertaken to analyze the damage and to recommend repairs and seismic upgrading. Subsequently, the owner, the General Services Administration, requested a consultant to conduct an additional study and examine the feasibility of seismic isolation as an alternative retrofit measure.

The study concluded that the damage sustained has significantly weakened the seismic resistance of the building and developed alternative retrofitting schemes. The conventional retrofitting schemes called for the addition of either shear walls or shear wall towers and general strengthening of the highly ornamented partition walls. The seismic isolation retrofitting schemes consisted of installation of isolation bearings and addition of shear walls.

The consultant recommended the use of seismic isolation despite an estimated moderate additional cost compared to the other retrofitting schemes. The recommendation was based on the following considerations: (a) greater degree of seismic protection, (b) least impact on its historic and architectural character, and (c) cost of repairing non-structural elements that may be damaged in a future severe earthquake.

The General Services Administration decided on the use of seismic isolation and in July of 1991 requested proposals on three isolation systems: high damping rubber bearings, lead-rubber bearings and friction pendulum system (FPS) bearings. Following an evaluation by a panel, which was based on technical merit, cost and experience of the supplier, the friction pendulum system was selected (Amin 1993).

CONCLUDING REMARKS

As Tables 1 to 4 show, significant strides have been made in the U.S. in the implementation of seismic isolation technology. This is a result of recent advances in seismic isolation hardware and the removal of several barriers, as discussed herein which had impeded its progress. A noteworthy feature associated with the U.S. practice is that isolation technology has been used for retrofitting buildings to a significant degree and a majority of these projects have been sponsored by the Federal or State governments.

REFERENCES

1.   American Association of State Highway and Transportation Officials
      AASHTO (1991).   "Guide specifications for seismic isolation design."
      Washington, D.C.

2.   Amin, N., Mokha, A., and Fatehi, H. (1993). "Seismic isolation retrofit of the
      U.S. Court of Appeals building."   Proc. ATC-17-1 seminar on seismic
      isolation, passive energy dissipation, and active control, March, San Francisco,
      CA.

3.   Buckle, I. G. and Mayes, R.L. (1990). "Seismic isolation history, application,
      and performance-- a world view." Earthquake Spectra, 6(2), 161-201.

4.   Constantinou, M.C., Kartoum, A., Reinhorn, A.M. and Bradford, P. (1991a).
      "Experimental and theoretical study of a sliding isolation system for bridges."
      Report No. NCEER 91-0027, National Center for Earthquake Engineering
      Research, Buffalo, NY.

5.   Constantinou, M.C. Mokha, A., and Reinhorn, A.M. (1991b). "Study of sliding
      bearing and helical-steel-spring isolation system." J. Struct. Engrg., ASCE,
      117(4), 1257-1275.

6.   Constantinou, M.C. (1992a). "NCEER-Taisei research on sliding isolation
      ssytems for bridges." NCEER Bulletin, 6(3), 1-4, July.

7.   Constantinou, M.C., Fujii, S., Tsopelas, P. and Okamoto, S. (1992b).
      "University at Buffalo-Taisei Corporation research project on bridge seismic
      isolation systems." 3rd Workshop on Bridge Engineering Research in Progress,
      La Jolla, CA, November.

8.   Constantinou, M.C., Symans, M.D., Tsopelas, P. and Taylor, D.P. (1993a).
      "Fluid viscous dampers in applications of seismic energy dissipation and
      seismic isolation." Proc. ATC-17-1 Seminar on Seismic Isolation, Passive
      Energy Dissipation, and Active Control, San Francisco, March.

9.   Constantinou, M.C., Tsopelas, P., Kim, Y-S. and Okamoto, S. (1993b).
      "NCEER-TAISEI Corporation research program on sliding seismic isolation
      systems for bridges - experimental and analytical study of friction pendulum
      system (FPS), "Report No. 93-0020, National Center for Earthquake
      Engineering Research, Buffalo, N.Y.

10. Constantinou, M.C., Tsopelas, P. and Okamoto, S. (1994). "Experimental of a class of bridge sliding seismic isolation systems," 3rd U.S. -Japan Workshop on Earthquake Protective systems for Bridges, Berkeley, CA, January.

11. Demetriades, G.F., Constantinou, M.C. and Reinhorn, A.M. (1992). "Study of wire rope systems for seismic protection of equipment in buildings." Report No. NCEER-92-0012, National Center for Earthquake Engineering Research, Buffalo, NY.

12. International Conference of Building Officials (1991). "Uniform building code, earthquake regulations for seismic-isolated structures. Whittier, CA.

13. Kelly, J.M., Beucke, K.E. and Skinner, M.S. (1980a). "Experimental testing of a friction damped aseismic base isolation system with fail-safe characteristics." Report No. UCB/EERC-80/18, University of California, Berkeley.

16. Kelly, J.M., Skinner, M.S. and Beuke, K.E. (1980b). "Experimental testing of an energy-absorbing base isolation system," Report No. UCB/EERC-80/35, University of California, Berkeley.

17. Kelly, J.M. and Hodder, S.B. (1981). "Experimental study of lead and elastomeric dampers for base isolation systems." Report No. UCB/EERC-81/16, University of California, Berkeley.

18. Kelly, J.M. (1982). "The influence of base isolation on the seimsic response of light secondary equipment." Report No. UCB/EERC-81/17, University of California, Berkeley.

19. Kelly, J.M. and Chalhoub, M.S. (1990). "Earthquake simulator testing of a combined sliding bearing and rubber bearing isolation system." Report No. UCB/EERC-87/04, University of California, Berkeley.

20. Kelly, J.M. (1991). "Base isolation: origins and development". EERC News, 12(1), 1-3, January.

21. Kelly, J.M. (1993). "State-of-the-art and state-of-the-practice in base isolation." Proc. ATC-17-1 seminar on seismic isolation, passive energy dissipation, and active control, March, San Francsco, CA.

22. Kircher, C.A., Delfosse, G.C., Schoof, C.C., Khemici, O. and Shah, H.C. (1979). "Performance of a 230 kV ATB7 power circuit breaker mounted on Gapec seismic isolators." Report No. 40, J.A. Blume Earthquake Engineering Center, Stanford University, September.

23.   Makris, N. and Constantinou, M.C. (1992). "Spring-viscous damper systems for combined seismic and vibration isolation." Earthquake Engineering and Structural Dynamics, 21(8), 649-664.

24.   Mayes, R.L., Sveinsson, B.I., and Buckle, I.G. (1988). "Analysis, design and testing of the isolation system for the Salt Lake City and county building: Proc. International Seismic Isolation/Historic Preservation Symposium, Salt Lake City, UT.

25.   Mayes, R.L., Jones, L.R. and Buckle, I.G. (1990). "Impediments to the implementation of seismic isolation." Earthquake Spectra, 6(2), 283-296.

26.   Mokha, A., Constantinou, M.C. and Reinhorn, A.M. (1988). "Teflon bearings in aseismic base isolation: Experimental studies and mathematical modeling, "Report No. NCEER-88-0038, National Center for Earthquake Engineering Research, Buffalo, NY.

27.   Mokha, A., Constantinou, M.C., Reinhorn, A.M., and Zayas, V. (1991). "Experimental study of friction pendulum isolation system." J. Struct. Engng., ASCE, 117(4), 1201-1217.

28.   Nagarajaiah, S., Reinhorn, A.M., and Constantinou, M.C. (1991). "3D-BASIS: Nonlinear dynamic analysis of three-dimensional base isolated structures." Report No. NCEER-91-0005, National Center for Earthquake Engineering Research, Buffalo, NY.

29.   Tsopelas, P.C., Nagarajaiah, S., Constantinou, M.C., and Reinhorn, A.M. (1991). "Nonlinear dynamic analysis of multiple building base isolated structures. Program 3D-BASIS-M." Report No. NCEER-91-0014, National Center for Earthquake Engineering Research, Buffalo, NY.

30.   Tsopelas, P., Okamoto, S., Constantinou, M.C., Ozaki, D. and Fujii, S., (1994). "NCEER-TAISEI Corporation research program on sliding seismic isolation systems for bridges - experimental and analytical study of systems consisting of sliding bearings, rubber restoring force devices and fluid dampers." Report No. 94-0002, National Center for Earthquake Engineering Research, Buffalo, N.Y.

31.   Zayas, V., Low, S.S. and Mahin, S.A. (1987). "The FPS earthquake resisting system, experimental report." Report No. UCB/EERC-87-01, University of California, Berkeley.

Table 1 . Directory of Seismically Isolated Buildings in the United States

Structure	Isolation System	Construction Completed	Floors /Size (m^2)	Structural System
Foothill Communities, Law and Justice Center, San Bernardino, CA (new)	98 high-damping rubber bearings	1986	5/3280	Steel braced frame
Flight Simulator Manufacturing Facility, Salt Lake City, Utah (new)	40 lead-rubber + 58 rubber bearings	1988	5/17100	Steel moment frame
Salt Lake City and County Building, Salt Lake City, Utah (retrofit)	208 lead-rubber + 239 rubber bearings	1989	5/15800	URM bearing wall/steel braced frame for clock tower
Fire Command and Control Facility, Los Angeles, CA (new)	32 high-damping rubber bearings	1990	2/3040	Steel braced frame
Two Townhouses, Los Angeles, CA (new)	-Each- 17 spring units + 6 viscodamper units (seismic + vibration isolation)	1990	3/440 3/400	Steel braced frame
USC Teaching Hospital, Los Angeles, CA (new)	68 lead-rubber + 81 rubber bearings	1991	8/31800	Steel braced frame
Hawley Apartments, San Francisco, CA (retrofit)	31 friction pendulum system (FPS) bearings	1991	4/1860	Wood bearing wall/steel moment frame at 1st floor
Rockwell Building 80, Seal Beach, CA (retrofit)	52 lead-rubber + 26 rubber bearings	1991	8/23600	R/C moment frame
Kaiser Computer Center, Corona, CA (new)	54 lead-rubber + high damping rubber bearings	1992	2/11150	Steel frame
MacKay School of Mines, Reno, Nevada (retrofit)	64 high-damping rubber bearings + 42 PTFE sliders	1993	3/4650	URM bearing wall

Table 1 Directory of Seismically Isolated Buildings in the United States (continued)

Structure	Isolation System	Construction Completed	Floors/ Size (m²)	Structural System
US Court of Appeals, San Francisco, CA (retrofit)	256 friction pendulum system (FPS) bearings	retrofit in progress in 1994	5/31500	Steel frame/URM in-fill R/C shear wall
Emergency Operations Center, East Los Angeles, CA (new)	28 high damping rubber bearings (Bridgestone)	under construction in 1994	18/14200	Steel Frame
Oakland City Hall, Oakland, CA (retrofit)	42 lead-rubber + 69 rubber bearings	retrofit in progress in 1994	18/14200	Steel Frame
Portland Water Bureau, OR (new)	31 lead rubber + 4 rubber bearings	1993	2/2550	Steel Frame
San Francisco Main Library, CA (new)	47 lead rubber + 95 rubber bearings	under construction in 1994	7/34300	Steel Frame
Channing House Palo Alto, CA (retrofit)	56 lead-rubber + 58 rubber bearings	under construction in 1994	10/23600	R/C moment frame
MLK-Drew Trauma Center, Watts, CA (new)	70 high damping rubber bearings	under construction in 1994	6/15900	steel braced frame
VA Hospital Long Beach, CA (retrofit)	110 lead-rubber + 18 rubber bearings	under construction in 1994	12/31800	R/C wall & frame

Table 1 Directory of Seismically Isolated Buildings in the United States (continued)

Structure	Isolation System	Construction Completed	Floors/ Size (m^2)	Structural System
Kerckhoff Hall, UCLA, Los Angeles, CA (retrofit)	57 lead-rubber + 69 rubber bearings	under construction in 1994	6/11400	R/C frame, Ext. brick infill, limited shear walls
Hughes Aircraft, Los Angeles, CA (retrofit)	24 lead-rubber + 19 rubber bearings	under construction in 1994	12/21800	R/C wall & frame
Autozone HQ, Memphis, TN (new)	24 lead-rubber + 19 rubber bearings	under construction in 1994	8/19100	Shear wall-long., steel frame-trans.
Campbell Hall, Western Oregon State College, Menmouth, OR (new)	26 lead-rubber + 16 rubber bearings	under construction in 1994	3/1800	URM bearing wall
San Bernardino County Medical Center (new)	400 high damping rubber bearings + 233 nonlinear viscous fluid dampers	construction to begin in 1994	5 buildings, floor area = 84000	Steel frame
Educational Services Center, Los Angeles, CA (retrofit)	PTFE sliding bearings and restoring force devices retrofit	retrofit to begin in 1994	9/8360	R/C Frame
R/C: Reinforced Concrete, URM: Unreinforced Masonry				

Table 2 Isolated Bridges in the United States

State	Number of Bridges	Total Isolated Length(m)	Isolation System
Alabama	1 (new)	141	Lead-rubber bearings
California	2 (new) 6 (retrofit)	1307	Lead-rubber bearings
Connecticut	1 (retrofit)	369	Lead-rubber bearings
Illinois	4 (new) 3 (retrofit)	2068	Lead-rubber bearings Lubricated PTFE sliding bearings with yielding steel dampers
Indiana	1 (new) 1 (retrofit)	352	Lead-rubber bearings PTFE sliding bearings and restoring force devices
Kentucky	1 (new) 1 (retrofit)	584	Lead-rubber bearings
Massachusetts	3 (new) 1 (retrofit)	1683	Lead-rubber bearings
Missouri	7 (new)	989	Lead-rubber bearings
Nevada	1 (retrofit)	135	Lead-rubber bearings
New Hampshire	5 (new) 1 (retrofit)	720	Lead-rubber bearings PTFE sliding bearings and restoring force devices
New Jersey	2 (new) 3 (retrofit)	718	Lead-rubber bearings
New York	4 (retrofit) 1 (new)	703	Lead-rubber bearings
Oregon	1 (new) 1 (retrofit)	624	Lead-rubber bearings Lubricated PTFE sliding bearings with yielding steel dampers
Pennsylvania	1 (new)	54	Lead-rubber bearings

Table 2 Isolated Bridges in the United States (continued)

State	Number of Bridges	Total Isolated Length(m)	Isolation System
Rhode Island	1 (new) 1 (retrofit)	698	Lead-rubber bearings
Vermont	1 (retrofit)	77	Lead-rubber bearings
Washington	1 (new) 1 (retrofit)	503	Lead-rubber bearings
West Virgina	2 (new)	326	Lead-rubber bearings

Table 3 List of Other Isolated Structures in the United States

Isolated Structure	Isolated System	Year of Construction
Circuit Breakers, South California	Low damping rubber bearings	1979
Mark II Detector, Stanford Linear Accelerator, CA	Lead-rubber bearings	1987
Liquid Argon Calorimeter, Stanford Linear Accelerator, CA	Lead-rubber bearings	1987
Emergency Water Tank Dow Chemical, San Francisco	Friction pendulum system (FPS) bearings	1988
Titan Solid Rocket Motor Storage, Vandenburg Air Force Base, CA	High damping rubber bearings	1991
Ammonia Storage Tank, ISP Chemicals, Calvert City, Kentucky	Friction pendulum system (FPS) bearings	1993

Table 4　List of Seismic Isolation Retrofit Projects in the United States

Structure	Status	Original Structural System	New Structural System	Isolation System	Motivating Factors for Selecting Isolation	Owner
Salt Lake City and County Building, Salt Lake City, Utah	Completed in 1989	1894 5-story URM bearing wall with 67 m clock tower	Steel braced frame for clock tower, floor strengthening	208 lead rubber + 239 rubber bearings	Historical building preservation	Salt Lake City Corporation
Rockwell Building 80, Seal Beach, CA	Completed in 1991	1967 8-story R/C moment frame	R/C moment frame at perimeter and floor 1 to 6	52 lead rubber + 26 rubber bearings	Functionality	Rockwell International
Hawley Apartments, San Francisco, CA	Completed in 1991	1920 4-story wood bearing wall	Steel moment frame at 1st floor	31 friction pendulum system bearings	Design economy	Private
MacKay School of Mines, Reno, Nevada	Completed in 1992	1908 3-story URM bearing wall	Floor ties/wall anchors, new basemat	64 high-damping rubber bearings + 42 teflon sliders	Historical building preservation	University of Nevada, Reno
U.S. Court of Appeals, San Francisco, CA	Retrofit in progress in 1994	1905 4-story steel frame/URM in -fill with 1933 addition	R/C shear walls, limited floor strengthening	256 friction pendulum system bearings	Historical building preservation	General Services Administration
Long Beach VA Hospital, CA	Retrofit in progress in 1994	1967 12-story, R/C shear walls	Strengthening of basement columns	110 lead rubber bearings + 18 rubber bearings	Functionality	Veterans Administration
Channing House, Palo Alto, CA	Retrofit in progress in 1994	R/C frame, shear walls	None	56 lead rubber + 58 rubber bearings	Design economy, investment protection	Private
Kerckhoft Hall, UCLA, Los Angeles, CA	Retrofit in progress in 1994	R/C frame and URM	None	57 lead rubber + 69 rubber bearings	Investment protection	UCLA
Hughes Aircraft, Los Angeles, CA	Retrofit in progress in 1994	R/C frame	None	24 lead rubber + 21 rubber bearings	Investment protection	Hughes Corporation

Table 4 List of Seismic Isolation Retrofit Projects in the United States (continued)

Structure	Status	Original Structural System	New Structural System	Isolation System	Motivating Factors for Selecting Isolation	Owner
Campbell Hall, Western Oregon State College, Menmouth, OR	Retrofit in progress in 1994	URM bearing wall	None	26 lead rubber + 16 rubber bearing	Design economy, historical buildiing preservation	Western Oregon State College
Oakland City Hall, Oakland, Ca	Retrofit in progress in 1994	1914 steel frame/URM in-fill	steel braced frame for clock tower and office tower, R/C shear walls, trusses above isolators	42 lead rubber + 69 rubber bearings	Design economy, historical building preservation	City of Oakland
State of California Justice Building, San Francisco, Ca	Conceptual design completed in 1992, awaits funding	1916 6-story steel frame with R/C floors/brick in-fill	R/C shear walls	170 high damping rubber bearings	Investment protection	State of California
50 UN Plaza Building, San Francisco, CA	Conceptual design completed in 1992, awaits funding	1936 6-story steel frame with R/C floors/brick in-fill	R/C shear walls	Isolation system to be selected in competitive bidding	Investment protection, functionality	General Services Administration
San Francisco Asian Art Museum, San Francisco, CA	Conceptual design completed, awaits funding	1916 3-story steel frame/URM in-fill	R/C shear walls	Not yet decided (approximately 200 isolators)	Content protection	City of San Francisco

CHAPTER VIII

BASE ISOLATION DEVELOPMENT IN JAPAN
- CODE PROVISIONS AND IMPLEMENTATION -

H. Iemura
Kyoto University, Kyoto, Japan

ABSTRACT

In recent years, base isolation techniques are actively adopted in construction of bridges, buildings and other structures in Japan. This paper summarizes the code provisions for base isolated bridges and buildings in Japan, as well as their implementation. Earthquake records obtained from base isolated buildings are also reported to verify reduction of earthquake response in base-isolated structures.

Development of Menshin Design Manual of Highway Bridges in Japan

A three-year Joint Research Program between the Public Works Research Institute and 26 groups consisting of 28 private firms on "Development of Menshin Design of Highway Bridges" was conducted from April 1989 to March 1992. The objective of this program is to develop a rational seismic design method of highway bridges with energy dissipating devices. Highway bridges with span length from 20m to 100m are considered as the major target for this joint research program.

The term "menshin" means reduction of response in Japanese. Although menshin design is closely related to the seismic isolation design, natural period of bridge is not forcibly elongated in the menshin design, because there are various restrictions in increasing the natural period. Instead of elongating the period, emphasis in the menshin design is on increasing energy dissipating capability and distribution of lateral forces to as many substructures as possible in order to decrease the lateral forces for design of substructures.

In this program, concentrated efforts were made to 1) development of menshin (energy dissipating) devices for highway bridges, 2) development of falling-off prevention devices and expansion joints appropriate for the mention bridges, 3) development of rational and simple menshin design method, and 4) favorable application of menshin design to highway bridges.

In the research project, four high-damping rubber bearings, two sliding friction dampers, a steel damper, a roller menshin bearing, a link bearing, and a viscous damper were newly developed for bridges.

In menshin design, the menshin devices are designed by the Seismic Coefficient Method and the Bearing Capacity Method, In both method, the lateral force is statically applied to the bridge, and seismic safety is examined based on the allowable stress design approach in the Seismic Coefficient Method and bearing capacity basis considering ductility in the Bearing Capacity Method. Bridges are designed by the Seismic Coefficient Method, and then the ductility of reinforced concrete piers is checked by the Bearing Capacity Method.

In the Seismic Coefficient Method, the design lateral force coefficient k_h is given as

$$k_h = c_Z \cdot c_G \cdot c_I \cdot c_T \cdot c_E \cdot k_{h0} \geq 0.1$$

provided that $c_T \cdot c_E \geq 0.8$

where,
k_h : lateral force coefficient
c_Z : modification factor for zone
c_G : modification factor for ground condition
c_I : modification factor for importance
c_T : modification factor for structural response depending on natural period (Fig.1)
c_E : modification factor energy dissipation capability
  ($c_E = 1.0$ for h_{eq} less than 0.1, $c_E = 0.9$ for h_{eq} larger than 0.1)
k_{h0} : standard design horizontal seismic coefficient (=0.2)

Except for c_E, all factors are the ones specified in the Specifications of Design of Highway Bridges.

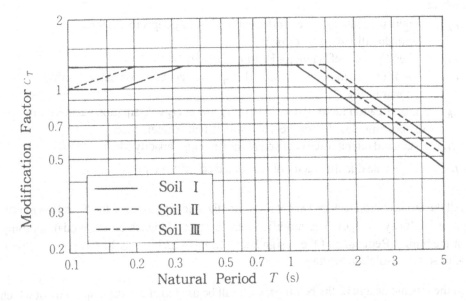

Fig. 1   Modification factor for structural response
depending on natural period of structure
(Seismic Coefficient Method)

The two significant mechanical properties of base isolation bearings to reduce earthquake response are 1) relatively soft stiffness for lateral load, and 2) high energy-damping capacity for repeated loads. Elongation of natural period of structures due to soft stiffness of base isolation bearings results in reduction of earthquake response. This effect can be accounted for in terms of the modification factor c_T in Fig. 1.

The modification coefficient c_E takes a value from 0.9 to 1.0 depending on the first modal damping ratio of the bridge. Reduction of the design lateral force by as large as 20 % is proposed in the Manual taking into account both the effects of low stiffness and high damping of base isolation bearings.

In the Bearing Capacity Method, the lateral force coefficient k_{h0} and the equivalent lateral force coefficient k_{he} are given as

$$k_{he} = \frac{k_{h0}}{\sqrt{2\mu - 1}} \geq 0.3$$

$$k_{h0} = c_Z \cdot c_I \cdot c_R \cdot c_E \cdot k_{h00}$$

where

k_{he} : equivalent lateral force coefficient for Bearing Capacity Method
with ductility demand of piers

k_{h0} : lateral force coefficient for Bearing Capacity Method

c_R : modification factor for structural response depending on natural period
shown in Fig. 2

c_E : modification factor (from 1.0 to 0.7) for energy dissipation capability due to
the bearing, ductility of piers and soil-structure interaction

k_{h00} : standard lateral force coefficient for Bearing Capacity Method (=1.0)

μ : allowable ductility factor of reinforced concrete piers

All the modification factors, except for c_E are the ones specified in the Specifications of Design of Highway Bridges. The modifications factor c_E depends on the modal damping ratio of the bridge. Reduction of the design lateral force by as large as 30% is proposed with the use of seismic isolation bearing.

In the seismic design of the bearings, k_{h0} shall be used to check the displacement which shall not be more than 250% of the height of rubber bearings, which guarantees stability of the bearings.

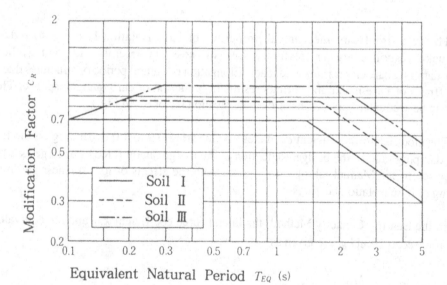

Fig. 2 Modification factor for structural response
depending on natural period of structure
(Bearing Capacity Method)

In the design of the piers, the reduced seismic coefficient k_{he} can be used with the given ductility demand μ.. However, k_{he} shall not be less than 0.3 to avoid extremely flexible structures.

Implementation of the Menshin Design

In Japan, several bridges have been built or are being built with seismic bearings of LRB or HDR type. The first bridge with the seismic isolation bearing in Japan is the Miyagawa Bridge in Shizuoka Prefecture. It is a three span continuous steel girder bridge of which seismic lateral force is distributed to two pies and two abutments with LRB. Its dimensions are shown in Fig.3. The Yamaage Bridge is a six-span continuous PC concrete girder bridge with HDR bearings of which the dimensions are shown in Fig.4. The free vibration and forced vibration tests were carried out to confirm fundamental vibration properties of the bridge. In addition to these two bridges, 8 bridges with seismic isolation bearings are being designed or built in Japan.

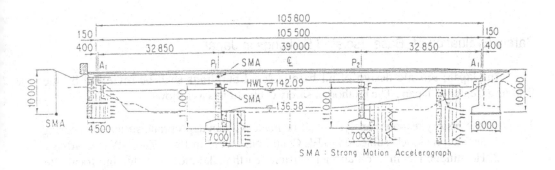

Fig.3   Side view of the Miyagawa Bridge

Due to the frequency of both small and large earthquakes in Japan, it is expected that the performance of these constructed bridges during actual earthquakes will give precious insight and knowledge on using seismic isolation for safety enhancement of bridges.

From the study on favorable application of menshin design, it has been found that the menshin is effective for constructing super-multispan continuous bridge with total deck length over 1 km. Application of the menshin design for seismic retrofit of existing bridges and deck connection for making existing simply supported girder bridges continuous were also studied and have been found feasible.

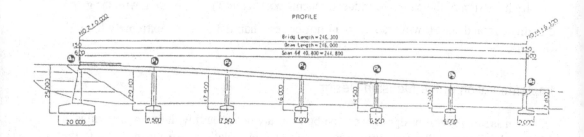

Fig.4 Side view of the Yamaage Bridge

Safety Evaluation of Base- Isolated Buildings in Japan

In Japan, seismic design of all base-isolated buildings shall be reviewed by the Building Center of Japan . The method of safety evaluation is as follows:

1. The time history response analysis shall be made with an appropriate structural, bearing and foundation models subjected to EL Centro NS record in 1940, Taft EW Record in 1 952, Hachinohe NS in 1968 and an artificial earthquake motion reflecting local site ground conditions.

2. Maximum acceleration of each earthquake ground motion is adjusted to give maximum velocity of 25 cm/sec and 50 cm/sec which corresponds to Level 1 and Level 2 earthquake motion.

3. For Level 1 earthquake ground motion, all structural members shall be under their yielding limits, and bearings should be under fairly stable region.

4. For Level 2 earthquake ground motion, seismic safety of base-isolated buildings is evaluated from behavior of both structures and bearings as shown in Table 1. It is strongly recommended that safety level shall be higher that B-.

Table 1 Safety Level of Base Isolated Buildings Subjected to Level 2 Earthquake Motion

Bearing`s Behavior / Structures` Behavior	B- I Bearing Displacement is within fairly Stable Range	B- II Bearing Displacement is over Stable Range but within Proof Tested Range
S- I All members are within Ultimate State,so interstory restoring force is within elastic limit	A+	B+
S- II Some members are over Ultimate State but interstory restoring force is within yielding limit	A−	B−
S- III Over S- II Range but Structural Stability is guaranteed	C	C

Implementation of Base-Isolated Buildings

For base-isolated buildings in Japan, the following types of bearing have been used.

Rubber bearings + Energy-Absorbing Damper

- Rubber Bearings + Inelastic Steel Damper

- Rubber Bearings + Inelastic Steel Damper + Lead Damper

- Rubber Bearings + Lead Damper + Friction Damper

- Rubber Bearings + Viscous Damper

Lead Rubber Bearing

High-Damping Rubber Bearing

Sliding Bearing

Construction of base-isolated buildings in Japan in each year is summarized in Table 2. It is found that rubber bearing + inelastic steel damper, lead rubber bearing , high-damper rubber bearing are popularly used for buildings.

Seismic Records Obtained at Base-Isolated Buildings

To verify the effectiveness of base isolation in reducing the earthquake response of structures, owners of the buildings are requested to install seismographs to record not only earthquake ground motion, but also earthquake response of base-isolated buildings.

On the campus of the Tohoku University in Miyagi Prefecture, two identical buildings, one with base-isolation bearings and the other with conventional foundation, were built to monitor actual behavior of existing buildings. Fig.5 shows the side view of the two buildings with and without the base isolation bearings. Obtained earthquake records in the two buildings are compared in Fig.6 which clearly verifies the effectiveness of the bearings.

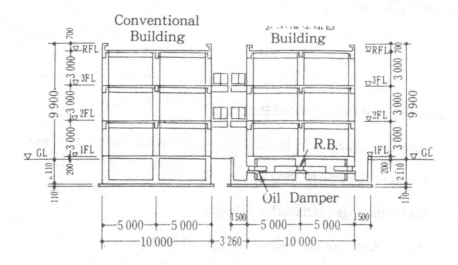

Fig.5 View of buildings with and without base isolation
(on the campus of Tohoku University)

Table 2 Base Isolated Buildings in Japan (Reviewed by Japan Center of Buildings)

| | Rubber Bearing | | | | Lead Rubber Bearing | High Damping Rubber Bearing | Sliding Bearing | Σ |
	Steel Damper	Steel Damper / Lead Damper	Lead Damper / Friction Damper	Viscous Damper				
1986	3	0	0	0	1	0	0	4
1987	4	0	0	1	1	0	0	6
1988	4	0	0	0	2	0	1	7
1989	4	1	0	1	2	4	1	13
1990	2	1	1	2	7	3	1	17
1991	3 (*1)	1	3 (*2)	0	3	1 (*3)	0	11
1992	2	0	1	0	3	3 (*4)	0	9
Σ	22	3	5	4	19	11	3	67

*1 One case is with oil damper

*2 Two cases are only with lead damper, one case is only friction damper

*3 Steel damper is also used

*4 One case is silicon rubber bearing, another case is with steel damper

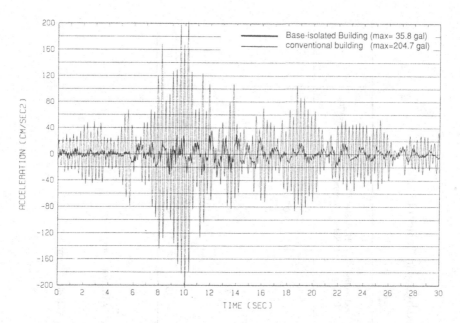

Fig.6 Maximum acceleration response at the top
of the buildings with and without base isolation

Acknowledgements

The author would like to thank Dr. William Tanzo, Research Associate at Saitama University for his help in preparation of this lecture notes.

References

Kawashima, K. (1992). "Manual for Menshin Design of Highway Bridges," 2nd U.S.-JapanWorkshop on Earthquake Protection Systems for Bridges.

Japan Road Association. (1990). Specifications for the Design of Highway Bridges, Part V-Seismic Design.February. (in Japanese)

Matsuo, Y.and Hara, K. (1991). "Design and Construction of Miyagawa Bridge (First Menshin Bridge in Japan)," 1st U.S.-Japan Workshop on Earthquake Protective Systems for Bridges, Buffalo, New York.

Iizuka,T., Kawakami, K., Kumakura, K. and Tani, H. (1992). "Menshin Design and construction of multi-span continuous prestressed concrete bridge," 2nd U.S.-Japan Workshop on Earthquake Protection Systems for Bridges.

Kubota, N., Shikauchi, Osaki, H., Mitake, Y. and Arai, H.(1992). "Deign of 9-span continuous prestressed concrete girder with base isolators," 2nd U.S.-Japan Workshop on Earthquake Protection Systems for Bridges.

Architectural Institute of Japan (1989). Recommendations for the Design of Base-Isolated Buildings.

CHAPTER IX

SEISMIC ISOLATION DEVELOPMENT IN EUROPE

M.C. Constantinou
State University of New York at Buffalo, Buffalo, NY, USA

ABSTRACT

This section presents a brief account of the development and application of seismic (or base) isolation in Europe. It appears that the first documented attempts to mitigate seismic hazard by seismic isolation were made in Europe.

INTRODUCTION

The first documented attempts to mitigate earthquake hazards by seismic isolation were made by Europeans. Bechtold of Munich, Germany obtained a U.S. patent in 1907 and Calantarients (an Armenian) of Scarborough, England applied for a British patent in 1909 (Kelly 1986, Buckle 1990). Both systems were primitive sliding isolation systems without restoring force.

In Italy following the devastating earthquake of 1908 in Messina-Reggio, a commission was formed and charged with the development of structural engineering methods for the reconstruction. Two proposals arose: one suggesting the use of sliding foundation, the other recommending the use of fixed foundation and requiring that buildings are designed for 8% of their weight as lateral force. The latter proposal was finally adopted and incorporated in building codes. The first application of seismic isolation came much later in 1959 and not in Europe. A structure was built in Ashkhabad, Turkmenistan by Russian engineers. The structure is suspended by cables in a simple but ingenious way so that it acts as a pendulum (Eisenberg 1992).

The documented simple ideas of Bechtold and Calantarients for seismic isolation have been actually successfully used much earlier. In the past, builders have used reed mats under building foundations to absorb seismic shocks. A notable example of such structure is a tall minaret in Kunya-Urgench, Turkmenistan, which was built in 1320 and is still standing.

The Greeks built the Parthenon in approximately 440 B. C. with its marble columns consisting of parts which were connected together by wood dowels. The dowels were covered with lead in an apparent attempt to extend their lifetime. The columns could rock at their joints and dissipate energy through friction and inelastic action in the lead. In effect Parthenon had the basic elements of an isolation system, that is, flexibility due to rocking of its columns and energy dissipation at the joints. It successfully withstood two thousand years of seismic activity until its partial destruction by an explosion in 1687.

DEVELOPMENT OF SEISMIC ISOLATION IN EUROPE

Laminated natural rubber bearings were first used in the U.K. at the Pelham Bridge in Lincoln in 1956. Over the next decade about 200 bridges were constructed in the U.K. on rubber bearings, all for accommodating thermal expansion.

Research by the Malaysian Rubber Producers Research Association (MRPRA) in the U.K. has led to the development of the high damping natural rubber. Testing of this new rubber compound was conducted at the University of California, Berkeley and it was first applied at the Foothill Communities building in California (Derham 1982, Kelly 1991). High damping rubber bearings are now manufactured in Italy, where they found several recent applications in the isolation of buildings (Martelli 1993).

European engineers proposed a variety of isolation systems and actually applied a number of them. A characteristic of some of these systems is that they were well conceived and fully functional. However, their application was not preceded by systematic and extensive research as done in the U.S. and Japan. The reason for this has been the lack of large scale testing facilities as those available in the U.S. since the late 1960s. Exemption has been the French EDF system which was extensively studied prior to its application. A number of isolation systems proposed in Europe are presented next.

Full Base Isolation or Seismafloat System

This system was proposed by Swiss engineers and patented by Seisma AG of Zurich. The isolated structure is supported by natural rubber blocks (without reinforcing steel plates) so that it provides a degree of flexibility in all directions. Under typical conditions a structure supported by such bearings will have a fundamental frequency in the translational-rocking mode of about 0.5 Hz and a vertical mode frequency of about 1.5 Hz. The system is enhanced by sacrificial elements (made of foam glass) which prevent motion in wind but break and allow motion in earthquakes. These elements break in a rather smooth fashion which, unlike steel shear pins, introduce only limited high frequency response in the superstructure. Testing of the system was conducted much later. A small 1/30 scale model was tested at ETH Lausanne. In 1982, a 1/3 scale, 36 kN model was tested at the University of California, Berkeley (Staudacher 1985). The system has been applied at the Pestalozzi school in Skopje in the former Yugoslavia in 1969. This 3-story, 24 MN weight structure was supported by 54 rubber blocks. It was designed by Swiss engineers (Staudacher 1985).

Earthquake Guarding System (Alexisismon)

Ikonomou (1972, 1984) in Greece proposed laminated elastomeric bearings and a combination of sliding bearings and rubber restoring force devices as two seismic isolation systems. The sliding isolation system (Figure 1) consists of multidirectional sliding pot bearings to carry the weight, provide the isolation mechanism and dissipate energy. Restoring force is provided by separate cylindrical rubber springs. It is a well conceived and functional system. However, the assumed values of friction coefficient in the sliding bearings were of the order of 0.02 to 0.03, which are typical of values obtained in very slow motions such as in the thermal expansion of bridges. Today we know that friction at high velocity of sliding, as in seismic excitation, is much higher and typically in the range of 0.08 to 0.12 for PTFE bearings, except for high bearing capacity composites for which friction may be as low as 0.05 (see lecture notes on "Properties of Sliding Bearings: Theory and Experiments"). The system found application in a structure in Athens, Greece.

Spring-Viscodamper Damper

Huffmann (1985) described the German GERB isolation system which consists of helical springs to support the weight of the isolated structure and provide the isolation mechanism in all directions. Viscodampers, consisting of a piston immersed in highly viscous fluid, provide the needed mechanism for energy dissipation.

This system was developed and extensively applied for the vibration isolation of structures, equipment and machinery. The viscodampers were developed in 1937 for the German navy for use in the isolation system of on board spring mounted Diesel engines.

Testing of this seismic isolation system was conducted at the Earthquake Research Institute at Skopje in the former Yugoslavia using a 5-story, 32-ton model structure (Huffmann 1985). The tests demonstrated the effectiveness of the system. Figure 2 shows a detail of the isolation system of the tested structure (from Huffmann 1985). It was assumed in the past that viscodampers behave as linear viscous dampers. In reality viscodampers exhibit strong viscoelastic behavior. The mechanical properties of viscodampers have been recently studied at the University at Buffalo. Experimental and analytical studies resulted in the development of analytical models which could predict the properties of these devices from their geometry and the mechanical properties of the fluid (Makris 1991, 1993). Figure 3 shows the mechanical properties of one such device. It may be seen that the damping coefficient reduces significantly with increasing frequency. This important property allows for the system to perform both as a seismic isolation system (low frequency response so that damping is high) and as a vibration isolation system (high frequency response so that damping is low) (Makris 1992).

The system found application in a large diesel generator in Taiwan in 1970 and very recently in two residential buildings in Los Angeles (Makris 1992).

GAPEC System

Delfosse, 1977 described a simple elastomeric isolation system which consists of low damping natural rubber bearings. The system found application in a school building, a small nuclear waste facility and four houses in France. The GAPEC isolators were also used in the isolation of a number of circuit breakers in California.

EDF System

This system was developed in France for use in nuclear power plants. In areas of low seismicity the system utilizes only neoprene bearings to produce limited isolation effect. In high seismicity areas, the neoprene bearings are fitted with top sliding plates which limit the transmission of force to about 20% of the carried weight (Plichon 1980, Pavot 1979, Gueraud 1985). An extended description of the system has been presented in the lecture notes on "Design and Applications of Sliding Bearings."

The system has been applied to two nuclear power plants in France and one in South Africa.

Sliding Bridge Isolation Systems in Italy

Italian engineers developed a number of sliding isolation systems for bridges. Typically, these systems utilize lubricated PTFE bearings together with energy dissipating devices such as yielding steel dampers or fluid dampers (Medeot 1991). A detailed description for one of these systems has been presented in the lecture notes on "Design and Applications of Sliding Bearings."

Approximately 150 bridges in Italy employ some form of these isolation systems (Medeot 1991, Martelli 1993). The total length of these isolated bridges is approximately 150 km.

SEISMIC ISOLATED STRUCTURES IN EUROPE

Table 1 presents a directory of seismically isolated structures in Europe. Italy has by far the largest number of applications. Currently, several more projects are in the design stage in Italy and Greece. Particularly, two 70m diameter LNG tanks in Greece will be constructed on an isolation system.

REFERENCES

1.   Buckle, I. G. and Mayes, R.L. (1990). "Seismic isolation history, application, and performance-- a world view." Earthquake Spectra, 6(2), 161-201.

2.   Delfosse, G.C. (1977). "The Gapec system: a new highly effective seismic system." Proc. 6th WCEE, 3, 1135-1140, New Delhi, India.

3.   Derham, C.J. (1982). "Basic principles of base isolation." Proc. International Conference on Natural Rubber for Earthquake Protection of Buildings and Vibration Isolation," 65-81, Kuala Lumpur, Malaysia.

4.   Eisenberg, J.M., Melentyev, A.M., Smirnov, V.I. and Nemykin, A.N. (1992). "Applications of seismic isolation in the USSR," Proc. 10th WCEE, Vol. 4, 2039-2046, Madrid.

5.   Guéraud, R., Noel-Leroux, J. P., Livolant, M. and Michalopoulos, A.P. (1985). "Seismic isolation using sliding-elastomer bearing pads," Nuclear Engrg. and Design, 84, 363-377.

6.   Huffmann, G.K. (1985). "Full base isolation for earthquake protection by helical springs and viscodampers." Nuclear Engineering and Design, 84, 331-338.

7.   Ikonomou, A.S. (1972). "The Earthquake guarding system." Technica Chronica, 879-898, October (in Greek and English).

8.   Ikonomou, A.S. (1984). "Alexisismon seismic isolation levels for translational and rotational seismic input." Proc. 8th WCEE, V, 957-982, San Francisco.

9.   Kelly, J.M. (1986). "Seismic base isolation: review and bibliography." Soil Dynamics and Earthquake Engineering, 5(3), 202-216.

10.   Kelly, J.M. (1991). "Base isolation: origins and development." EERC News, 12(1), 1-3.

11.   Makris, N. and Constantinou, M.C. (1991). "Fractional derivative Maxwell model for viscous dampers." Journal of Structural Engineering, ASCE, 117(9), 2708-2724.

12.   Makris, N. and Constantinou, M.C. (1992). "Spring-viscous damper systems for combined seismic and vibration isolation." Earthquake Engineering and Structural Dynamics, 21(8), 649-664.

13.   Makris, N. Constantinou, M.C. and Dargush, G. (1993). "Analytical model of viscoelastic fluid dampers." Journal of Structural Engineering, ASCE, 119 (11), 3310-3325.

14.   Martelli, A., Parducci, A. and Forni, M. (1993). "Innovative seismic design techniques in Italy." Proc. ATC-17-1 Seminar on Seismic Isolation, Passive Energy Dissipation, and Active Control, March, San Francisco, CA.

15.   Medeot, R. (1991). "The evolution of seismic devices for bridges in Italy." 3rd World Congress on Joint Sealing and Bearing Systems for Concrete Structures, Vol. 2, 1295-1320, Toronto.

16.   Pavot, B. and Polust, E. (1979). "Seismic bearing pads." Tribology International, 12(3), 107-111.

17.   Plichon, C., Guéraud, R., Richli, M.H. and Casagrande, J.F. (1980). "Protection of nuclear power plants against seism." Nuclear Technology, 49, 295-306.

18.   Staudacher, K. (1985). "Protection for structures in extreme earthquakes: full base isolation (3-D) by the Swiss Seismafloat system." Nuclear Engineering and Design, 84, 343-357.

Table 1 Directory of Isolated Structures in Europe

Country	Structure	Isolation System	Year
England	Nuclear Fuel Processing Facility	Elastomeric Bearings	N.A.
France	Cruas Nuclear Plant	Neoprene Bearings (EDF System without Sliding Part)	1981
	Le Pellirin Nuclear Plant	Neoprene Bearings (EDF System without Sliding Part)	1983
	3-story School in Lambesc	152 Low Damping Elastomeric Bearings (GAPEC)	1978
	Nuclear Waste Storage Facility	32 Low Damping Elastomeric Bearings (GAPEC)	1982
	4 houses	Low Damping Elastomeric Bearing (GAPEC)	1977-1982
Greece	Justice Building, Athens	PTFE Sliding Bearings and Restoring Force Devices	Construction began in 1973, then halted and completed at a later time.
	Two 70m Diameter LNG Storage Tanks	Friction Pendulum (FPS) Bearings	Under Construction in 1994.

Table 1 Directory of Isolated Structures in Europe (continued)

Country	Structure	Isolation System	Year
Iceland	5 bridges	Lead-rubber Bearings	N.A.
Italy	Fire Station, Napoli	24 Neoprene Bearings	1983
	SIP Center, Ancona (5 buildings)	297 High Damping Rubber Bearings	1989-1992
	Apartment Building, Squillace Marina	43 High Damping Rubber Bearings	1992
	Navy Building, Ancona	44 High Damping Rubber Bearings	1992
	Apartment houses of the Italian Navy, Campo Palma	192 High Damping Rubber Bearings	1993
	156 bridges of total length = 150 km	Sliding Bearings Rubber Bearings Lead-rubber Bearings Various Hysteretic Damping Devices	1974-1993
Former Yugoslavia	Pestalozzi School, Skopje	54 Natural Rubber Blocks (Seismafloat System)	1969

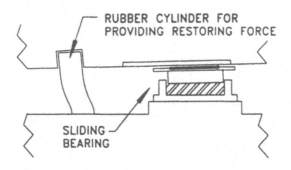

Figure 1   Alexisismon Sliding Isolation System

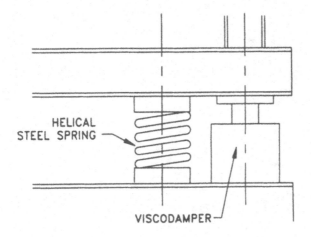

Figure 2   Components of GERB Spring-Viscodamper Isolation System

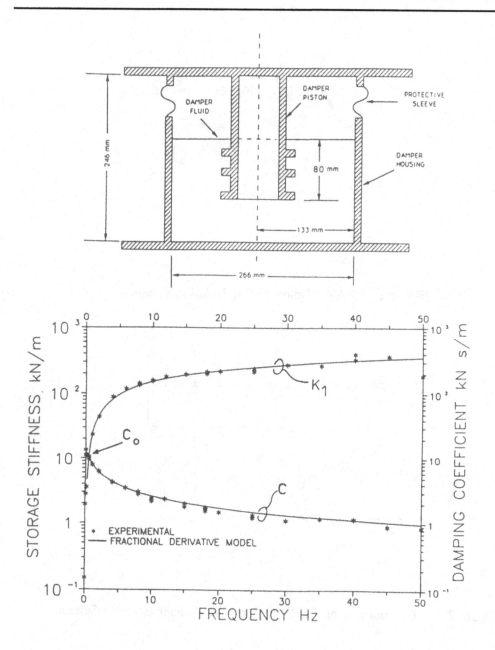

Figure 3   Mechanical Properties (K_1 = storage stiffness, C = damping coefficient) of Viscodamper.

CHAPTER X

PRINCIPLES OF FRICTION, VISCOELASTIC,
YIELDING STEEL AND FLUID VISCOUS DAMPERS:
PROPERTIES AND DESIGN

M.C. Constantinou
State University of New York at Buffalo, Buffalo, NY, USA

ABSTRACT

This section describes passive energy dissipating devices which may be used within a structural system to absorb seismic energy. These devices are capable of producing significant reductions of interstory drifts in moment-resisting frames. Furthermore, these devices may, under elastic conditions, reduce the design forces.

INTRODUCTION

The mitigation of the hazardous effects of earthquakes begins with the consideration of the distribution of energy within a structure. During a seismic event, a finite quantity of energy is input into a structure. This input energy is transformed into both kinetic and potential (strain) energy which must be either absorbed or dissipated through heat. If there were no damping, vibrations would exist for all time. However, there is always some level of inherent damping which withdraws energy from the system and therefore reduces the amplitude of vibration until the motion ceases. The structural performance can be improved if a portion of the input energy can be absorbed, not by the structure itself, but by some type of supplemental "device". This is made clear by considering the conservation of energy relationship (Uang 1988):

$$E = E_k + E_s + E_h + E_d \tag{1}$$

where E is the absolute energy input from the earthquake motion, E_k is the absolute kinetic energy, E_s is the recoverable elastic strain energy, E_h is the irrecoverable energy dissipated by the structural system through inelastic or other forms of action, and E_d is the energy dissipated by supplemental damping devices. The absolute energy input, E, represents the work done by the total base shear force at the foundation on the ground (foundation) displacement. It, thus, contains the effect of the inertia forces of the structure.

In the conventional design approach, acceptable structural performance is accomplished by the occurrence of inelastic deformations. This has the direct effect of increasing energy F_h. It also has an indirect effect. The occurrence of inelastic deformations results in softening of the structural system which itself modifies the absolute input energy. In effect, the increased flexibility acts as a filter which reflects a portion of the earthquake energy.

The technique of seismic isolation accomplishes the same task by the introduction, at the foundation of a structure, of a system which is characterized by flexibility and energy absorption capability. The flexibility alone, typically expressed by a period of the order of 2 seconds, is sufficient to reflect a major portion of the earthquake energy so that inelastic action does not occur. Energy dissipation in the isolation system is then useful in limiting the displacement response and in avoiding resonances. However, in earthquakes rich in long period components, it is not possible to provide sufficient flexibility for the reflection of the earthquake energy. In this case, energy absorption plays an important role.

Modern seismic isolation systems incorporate energy dissipating mechanisms. Examples are high damping elastomeric bearings, lead plugs in elastomeric bearings, mild steel dampers, fluid viscous dampers, and friction in sliding bearings (see lecture notes on Seismic Isolation Systems: Introduction and Overview).

Another approach to improved earthquake response performance and damage control is that of supplemental damping systems. In these systems, mechanical devices are incorporated in the frame of the structure and dissipate energy throughout the height of the structure. The means by which energy is dissipated is either yielding of mild steel, sliding friction, motion of a piston within a viscous fluid, orificing of fluid, or viscoelastic action in polymeric materials.

FRICTION DEVICES

A frictional device located at the intersection of cross bracing has been proposed by Pall (1982, 1987) and used in six buildings in Canada. These buildings are the Concordia University new library 10-story building in Montreal, the Canadian Space Agency building in St.-Hubert, the CCRIT building in Laval and three 3-story buildings of Ecole Polyvalante in Sorel which were damaged in the 1988 Saguenay earthquake (Pall 1993). Figure 1 illustrates the design of this device. When seismic load is applied, the compression brace buckles while the tension brace induces slippage at the friction joint. This, in turn, activates the four links which force the compression brace to slip. In this manner, energy is dissipated in both braces while they are designed to be effective in tension only.

Experimental studies by Filiatrault (1985) and Aiken (1988) confirmed that these friction devices could enhance the seismic performance of structures. The devices provided a substantial increase in energy dissipation capacity and reduced drifts in comparison to moment resisting frames. Reductions in story shear forces were moderate. However, these forces are primarily resisted by the braces in a controlled manner and only indirectly resisted by the primary structural elements. This subject is further discussed later.

Sumitomo Metal Industries of Japan developed, and for a number of years manufactured, friction dampers for railway applications. Recently, the application of these dampers was extended to structural engineering. Two tall structures in Japan, the Sonic City Office Building in Omiya City and the Asahi Beer Azumabashi Building in Tokyo, incorporate the Sumitomo friction dampers for reduction of the response to ground-borne vibrations and minor earthquakes. These structures are, respectively, 31- and 22-story steel frames. Furthermore, a 6-story seismically isolated building in Tokyo incorporates these dampers in the isolation system as energy-absorption devices.

Figure 2 shows the construction of a typical Sumitomo friction damper. The device consists of copper pads impregnated with graphite in contact with the steel casing of the device. The load on the contact surface is developed by a series of wedges which act under the compression of belleville washer springs. The graphite serves the purpose of lubricating the contact and ensuring a stable coefficient of friction and silent operation.

An experimental study of the Sumitomo damper was reported by Aiken (1990). Dampers were installed in a 9-story model structure and tested on a shake table. The dampers were not installed diagonally as braces. Rather, they were placed parallel to the floor beams, with one of their ends attached to a floor beam above and the other end attached to a chevron brace arrangement which was attached to the floor beam below. The chevron braces were designed to be very stiff. Furthermore, a special arrangement was used at the connection of each damper to the chevron brace to prevent lateral loading of the device. Figure 2 demonstrates the installation.

The experimental study resulted in conclusions which are similar to those of the study of the friction bracing devices of Pall (1982). In general, displacements were reduced in comparison to moment resisting frames. However, this reduction depended on the input motion. For example, in tests with the Japanese Miyagiken earthquake, ratios of interstory drift in the friction damped structure to interstory drift in the moment resisting structure of about 0.5 were recorded. In tests with the 1940 El Centro and 1952 Taft earthquakes, the ratio of interstory drifts was typically around 0.9. Furthermore, recorded base shear forces were, in general, of the same order as those of the moment resisting frame. However, the friction damped structure absorbed earthquake energy by mechanical means. This energy would have otherwise been absorbed by inelastic action in the frame.

An interesting outcome of the study is that, for optimum performance, the friction force at each level should be carefully selected based on the results of nonlinear dynamic analyses. The tested structure had a friction force of about $0.12W$ (W = model weight) at the first story and it reduced to about $0.05W$ at the top story.

Another friction device, proposed by Fitzgerald (1989), utilizes slotted bolted connections in concentrically braced connections. Component tests demonstrated stable frictional behavior. Figure 3 shows this friction device. It may be noted that the sliding interface is that of steel on steel.

Constantinou (1991a, 1991b) developed a friction device for application in bridge seismic isolation systems. Shown in Figure 4, this device utilizes an interface of stainless steel in contact with bronze which is impregnated with graphite. The device bears a similarity with the Sumitomo device in terms of the materials which form the sliding interface.

Very recently Grigorian (1993) tested a slotted bolted connection which was nearly identical to the one of Fitzgerard (1989) except for the sliding interface which consisted of brass in contact with steel. This interface exhibits more stable frictional characteristics than the steel on steel interface.

Frictional devices are in principle simple to construct but very difficult in maintaining their properties over prolonged time intervals. Particularly,

(a) Metal to metal interfaces typically promote additional corrosion. Specifically, common carbon and low alloy steels suffer severe additional corrosion when in contact with brass, bronze or copper in all atmospheric environments. The British Standards (BSI 1979) specifically recommend against their use. Only stainless steels with high content of chromium may be acceptable when in contact with brass or bronze under atmospheric conditions other than industrial/urban and marine (BSI 1979).

(b) The normal load on the sliding interface cannot be reliably maintained and some relaxation should be expected over the years.

YIELDING STEEL ELEMENTS

The reliable yielding properties of mild steel have been explored in a variety of ways for improving the seismic performance of structures. The eccentrically-braced frame (Roeder 1978) represents a widely accepted concept. Energy dissipation is primarily concentrated at specifically detailed shear links of eccentrically-braced frames. These links represent part of the structural system which is likely to suffer localized damage in severe earthquakes.

A number of mild steel devices have been developed in New Zealand (Tyler 1978, Skinner 1980). Some of these devices were tested at U.C. Berkeley as parts of seismic isolation systems (Kelly 1980) and similar ones were widely used in seismic isolation applications in Japan (Kelly 1988).

Tyler (1985) described tests on a steel element fabricated from round steel bar and incorporated in the bracing of frames. Figure 5 shows details of a similar bracing system which was installed in a building in New Zealand. An important characteristic of the element is that the compression brace disconnects from the rectangular steel frame so that buckling is prevented and pinched hysteretic behavior does not occur. Energy is dissipated by inelastic deformation of the rectangular steel frame in the diagonal direction of the tension brace.

Another element, called "Added Damping and Stiffness" or ADAS device has been studied by Whittaker (1989). The device consists of multiple X-steel plates of the shape shown in Figure 6 and installed as illustrated in the same figure. The similarity of the device to that of Tyler (1978) and Kelly (1980) is apparent. The shape of the device is such that yielding occurs over the entire length of the device. This is accomplished by the use of rigid boundary members so that the X-plates are deformed in double curvature.

Shake table tests of a 3-story steel model structure by Whittaker (1989) demonstrated that the ADAS elements improved the behavior of the moment-resisting frame to which they were installed by a) increasing its stiffness, b) increasing its strength and c) increasing its ability to dissipate energy. Ratios of recorded interstory drifts in the structure with ADAS elements to interstory drifts in the moment-resisting frame were typically in the range of 0.3 to 0.7. This reduction is primarily an effect of the increased stiffness of the structure by the ADAS elements.

Ratios of recorded base shears in the structure with ADAS elements to base shears in the moment-resisting frame were in the range of 0.6 to 1.25. Thus, the base shear in the ADAS frame was in some tests larger than the shear in the moment frame. However, it should be noted again that, as in the case of friction braced structures, the structure shear forces are primarily resisted by the ADAS elements and their supporting chevron braces (see Figure 6). The ADAS elements yield in a pre-determined manner and relieve the moment frame from excessive ductility demands. ADAS elements have been very recently used in the seismic retrofitting of the Wells Fargo Bank, a 2-story concrete building in San Francisco.

Various devices whose behavior is based on the yielding properties of mild steel have been implemented in Japan (Fujita 1991). Kajima Corporation developed bell-shaped steel devices which serve as added stiffness and damping elements. These dampers were installed in the connecting corridors between a 5-story and a 9-story building in Japan. The same company developed another steel device, called the Honeycomb Damper, for use as walls in buildings. They were installed in the 15-story Oujiseishi Headquarters Building in Tokyo. Obayashi Corporation developed a steel plate device which is installed in a manner similar to the ADAS elements (Figure 6). The plate is subjected to shearing action. It has been installed in the Sumitomo Irufine Office Building, a 14-story steel structure in Tokyo.

VISCOELASTIC DAMPERS

Viscoelastic dampers, made of bonded viscoelastic layers (acrylic polymers), have been developed by 3M Company and used in wind vibration control applications. Examples are the World Trade Center in New York City (110 stories), the Columbia SeaFirst Building in Seattle (73 stories) and the Number Two Union Square Building in Seattle (60 stories). Seismic applications of viscoelastic dampers have a more recent origin. For seismic applications, larger damping increases are usually required in comparison with those required for mitigation of wind-induced vibrations. Furthermore, energy input into the structure is usually spread over a wider frequency range, requiring more effective use of the viscoelastic materials. Extensive analytical and experimental studies in the seismic domain have led to the first seismic retrofit of an existing building using viscoelastic dampers in the U.S. in 1993.

Viscoelastic materials used in structural application dissipate energy when subjected to shear deformation. A typical viscoelastic (VE) damper is shown in Fig. 7 which consists of viscoelastic layers bonded to steel plates. When mounted in a structure, shear deformation and hence energy dissipation takes place when the structural vibration induces relative motion between the outer steel flanges and the center plate.

Under a sinusoidal load with frequency ω, the shear strain $\gamma(t)$ and the shear stress $\tau(t)$ oscillate at the same frequency ω but in general out-of-phase. They can be expressed by (Zhang 1989)

$$\gamma(t) = \gamma_o \sin \omega t, \quad \tau(t) = \tau_o \sin(\omega t + \delta) \tag{2}$$

where γ_o and τ_o are, respectively, the peak shear strain and peak shear stress, and δ is the lag angle. For a given γ_o, τ_o is a function of ω.

The shear stress can also be written as

$$\tau(t) = \gamma_o[G'(\omega)\sin\omega t + G''(\omega)\cos\omega t] \tag{3}$$

where $G'(\omega) = \tau_o\cos\delta/\gamma_o$ and $G''(\omega) = \tau_o\sin\delta/\gamma_o$. The stress-strain relationship can be written as

$$\tau(t) = G'(\omega)\gamma(t) \pm G''(\omega)\left[\gamma_o^2 - \gamma(t)\right]^{1/2} \tag{4}$$

which defines an ellipse as shown in Fig. 8, whose area gives the energy dissipated by the viscoelastic material per unit volume and per cycle of oscillation.

It is seen from Eq. (4) that the first term of the shear stress is the in-phase portion with $G'(\omega)$ representing the elastic stiffness, and the second term or the out-of-phase portion represents the energy dissipation component. This is seen more clearly if Eq. (3) is rewritten in the form

$$\tau(t) = G'(\omega)\gamma(t) + \frac{G''(\omega)}{\omega}\dot{\gamma}(t) \tag{5}$$

since $\dot{\gamma}(t) = \gamma_o\omega\cos\omega t$. The quantity $G''(\omega)/\omega$ is the damping coefficient of the damper material. The equivalent damping ratio is

$$\xi = \frac{G''(\omega)}{\omega}\left(\frac{\omega}{2G'(\omega)}\right) = \frac{G''(\omega)}{2G'(\omega)} \tag{6}$$

Accordingly, $G'(\omega)$ is defined as the *shear storage modulus* of the viscoelastic material, which is a measure of the energy stored and recovered per cycle; and $G''(\omega)$ is defined as the *shear loss modulus*, which gives a measure of the energy dissipated per cycle. The *loss factor*, η, defined by

$$\eta = \frac{G''(\omega)}{G'(\omega)} = \tan\delta \tag{7}$$

is also often used as a measure of energy dissipation capacity of the viscoelastic material.

The shear storage modulus and shear loss modulus of a viscoelastic material are generally functions of excitation frequency (ω), ambient temperature (T), shear strain (γ), and material temperature (θ). Their dependence on these parameters have been studied extensively both analytically and experimentally. Constitutive models that have been proposed for viscoelastic materials include the Maxwell model, the Kevin-Voight model and complex combinations of these elementary models. More recently, the concept of fractional derivatives (Tsai 1993, Kasai 1993) and constitutive modeling based on the Botzmann's superposition principle (Shen 1994) have been used in modeling VE damper properties.

In order to determine the dependence of $G'(\omega)$ and $G''(\omega)$ on the ambient temperature, the method of reduced variables can be used (Ferry 1980). This method affords a convenient simplification in separating the two principal variables, frequency and temperature, on which the VE material properties depend, and expressing these properties in terms of a single function of each, whose form can be experimentally determined.

The effect of temperature rise within the VE material when it is subjected to shear deformation can also be taken into account using the method of reduced variables. Analytically, the internal temperature, θ, within the VE material due to the mechanical work done by the damper can be calculated from the heat transfer equation (Kasai 1993)

$$s\rho \frac{\partial \theta}{\partial t} \cong K \frac{\partial^2 \theta}{\partial z^2} + \tau \frac{\partial \gamma}{\partial t} \tag{8}$$

where s is the specific heat of the VE material, ρ is the mass density, and K is thermal conductivity. The spatial variation of the temperature is assumed to occur in the z-direction across the VE layer. Experimental results and finite element analyses indicate, however, this transient heat conduction term is small and Eq. (8) can be approximated by

$$\theta(t) = T + \frac{1}{s\rho} \int_0^t \tau(t)\dot{\gamma}(t)dt \tag{9}$$

where T is the ambient temperature.

In structural applications, one is interested in the temperature rise within the VE material over a loading episode. Field observations and laboratory experiments have shown that, during each wind and earthquake loading cycle, this transient temperature increase is typically less than 10°C and has a minor effect on the performance of VE dampers.

In summary, assuming VE dampers undergo moderate strain ($\leq$ 20%), the shear storage and loss moduli can be considered as functions of only the excitation frequency ω and the ambient temperature T.

Based upon the development above, it is seen from Eq. (5) that, at a given ambient temperature and under moderate strain, the stress in a VE material is linearly related to the strain and strain rate under harmonic motion. For a viscoelastic damper such as that shown in Fig. 7 with total shear area A and total thickness h, the corresponding force-displacement relationship is

$$F(t) = k'(\omega)x(t) + c'(\omega)\dot{x}(t) \tag{10}$$

where

$$k'(\omega) = \frac{AG'(\omega)}{h}, \quad c'(\omega) = \frac{AG''(\omega)}{\omega h} \tag{11}$$

Thus, unlike friction devices or yielding steel elements, a linear structure with added VE dampers remains linear with the dampers contributing to increased viscous damping as well as lateral stiffness. This feature represents a significant simplification in the analysis of viscoelastically damped structures (Zhang 1989, 1992).

In order to assess seismic applicability of viscoelastic dampers, extensive experimental programs have been designed and carried out for steel frames in the laboratory (Ashour 1987, Su 1990, Lin 1991, Fujita 1992, Kirekawa 1992, Aiken 1993, Bergman 1993, Chang 1993a), for lightly reinforced concrete frames in the laboratory (Foutch 1993, Lobo 1993, Chang 1994), and for a full-scale steel frame structure in the field (Chang 1993a). These experimental results, together with analytical modeling, have led to the development of design procedures for structures with added VE dampers.

Like many other design problems, the design of viscoelastically damped structures is in general an iterative process (Chang 1993a and 1993b). First, an analysis of the structure without added dampers should be carried out. Then the required damping ratio becomes the primary design parameter for adding VE dampers to the structure. The design will normally contain the following steps which may continue to update the structural properties after each design cycle: (a) determine structural properties of the building and perform structural analysis; (b) determine the desired damping ratio; (c) select desirable and available damper locations in the building; (d) select damper stiffness and loss factor; (e) calculate the equivalent damping ratio using the modal strain energy method; and (f) perform structural analysis using the designed damping ratio. When steps (e) and (f) satisfy the desired damping ratio and the structural performance criteria, the design is complete. Otherwise, a new design cycle will proceed which may lead to new structural properties, damper locations or damper dimensions and properties.

It can be seen that this design procedure falls into the traditional design procedure except for the determination of the required damping ratio and the selection of damper stiffness and loss factor. In general, the required damping ratio can be estimated by using the response spectra of the design earthquake with various damping ratios. The selection of damper stiffness k' and loss factor η can be a trial and error procedure. They can also be determined based on the principle that the added stiffness due to the VE dampers be proportional to the story stiffness of the structure. This is obtained from

$$k'_i = \frac{2\zeta}{\eta - 2\zeta} k_i \qquad (12)$$

where ζ is the target damping ratio and k'_i and k_i are, respectively, the damper stiffness and the structural story stiffness without added dampers at the ith story. For a VE material with known G' and G'' at the design frequency and temperature, the area of the damper, A, can be determined from, as seen from the first of Eqs. (11),

$$A = \frac{k'h}{G'} \qquad (13)$$

The thickness of the VE material, h, can be determined from the maximum allowable damper deformation to insure that the maximum strain in the VE material is lower than the ultimate value.

In this design procedure, the structure is assumed to be linear elastic. If inelastic deformation is allowed in the structure, the demand in VE damping can be reduced and a modified procedure has to be used. We also note that ambient temperature is an important design parameter in all cases.

Viscoelastic devices have also been developed by the Lorant Group which may be used either at beam-column connections or as parts of a bracing system. Experimental and analytical studies have been very recently reported by Hsu (1992). These devices have been installed in a 2-story steel structure in Phoenix, Arizona.

Hazama Corporation of Japan developed a viscoelastic device whose construction and installation is similar to the 3M viscoelastic device with the exception that several layers of material are used (Fujita 1991). The material used in the Hazama device also exhibits temperature dependent properties. Typical results on the storage and loss shear moduli at a frequency of 1 Hz and shear strain of 0.5 are: 355 psi (2.45 MPa) and 412 psi (2.85 MPa), respectively at 32°F (0°C) and 14 psi (0.1 MPa) and 8 psi (0.055 MPa), respectively at 113°F (45°C).

Another viscoelastic device in the form of walls has been developed by Shimizu Corporation (Fujita 1991). The device consists of sheets of thermo-plastic rubber sandwiched between steel plates. It has been installed in the Shimizu Head Office Building, a 24-story structure in Tokyo.

VISCOUS WALLS

The Building Research Institute in Japan tested and installed viscous damping walls in a test structure for earthquake response observation. The walls were developed by Sumitomo Construction Company (Arima 1988) and consist of a moving plate within a highly viscous fluid which is contained within a wall container. The device exhibits strong viscoelastic fluid behavior which is similar to that of the GERB viscodampers which have been used in applications of vibration and seismic isolation (Makris 1992).

Observations of seismic response of a 4-story prototype building with viscous damping walls demonstrated a marked improvement in the response as compared to that of the building without the walls.

FLUID VISCOUS DAMPERS

Fluid viscous dampers which operate on the principle of fluid flow through orifices were first used in the 75mm French artillery rifle of 1897. This high performance damper was adaptive, that is, its output was continuously varied depending on the angle of elevation of the weapon. It was considered a national secret of France which was shared with the U.S. and Great Britain only during World War I. Even today elements of the design are used in large artillery pieces and in most aircraft landing gears.

Fluid dampers for automotive use were invented by Ralph Peo of Buffalo, N.Y. in 1925. Since then a number of fluid damper manufacturers were established in the Buffalo area. Of these, Taylor Devices was established in 1955 and since then produced over two million devices which were used as energy absorbing buffers in steel mills, canal lock buffers, offshore oil leg suspensions and primarily in shock and vibration isolation systems of aerospace and military hardware. Some notable examples of application are launch gantry dampers for the U.S. Navy with force output of up to 2000 kips (8900 kN) and travel of over 10 feet (over 3m), seismic dampers in nuclear power plants with force output of 300 to 1000 kips (1335 to 4450 kN), payload dampers for the space shuttle, wind dampers for the Atlas and Saturn V rockets and shock isolators for most tactical and strategic missiles of the U.S. Armed Forces.

A particular fluid damper which is produced by Taylor Devices has been studied by Constantinou (1992, 1993). The construction of this device is shown in Figure 9. It consists of a stainless steel piston with a bronze orifice head and an accumulator. It is filled with silicone oil. The orifice flow is compensated by a passive bi-metallic thermostat that allows operation of the device over a temperature range of - 40°F to 160°F (-40°C to 70°C). The orifice configuration, mechanical construction, fluid and thermostat used in this device originated within a device used in a classified application on the U.S. Air Force B-2 Stealth Bomber. Thus, the device includes performance characteristics considered as state of the art in hydraulic technology.

The force that is generated by the fluid damper is due to a pressure differential across the piston head. However, the fluid volume is reduced by the product of travel and piston rod area. Since the fluid is compressible, this reduction in fluid volume is accompanied by the development of a restoring (spring like) force. This is prevented by the use of the accumulator. Tested devices showed no measurable stiffness for piston motions with frequency less than about 4 Hz. In general, this cut-off frequency depends on the design of the accumulator and may be specified in the design. The existence of the aforementioned cut-off frequency is a desirable property. The devices may provide additional viscous type damping to the fundamental mode of the structure (typically with a frequency less than the cut-off frequency) and additional damping and stiffness to the higher modes. This may, in effect, completely suppress the contribution of the higher modes of vibration. Alternatively, fluid dampers may be constructed with run-through rod. This design prevents compression of the fluid and it does not require an accumulator.

The force in the fluid damper may be expressed as

$$P = C|\dot{u}|^{\alpha} sgn(\dot{u}) \tag{14}$$

where $\dot{u}$ = velocity of piston rod, C = constant and α = coefficient in the range of approximately 0.5 to 2.0. A typical orifice design features cylindrical orifices for which the output force is proportional to the velocity squared (α=2). This performance is usually unacceptable.

To produce damping forces with coefficient α different than 2 requires specially shaped passages to alter the flow characteristics with fluid speed. This orifice design is known as fluidic control orifice. A design with coefficient α equal to 0.5 is useful in applications involving extremely high velocity shocks. They are typically used in the shock isolation of military hardware. Fluid viscous dampers with nonlinear characteristics have been recently specified in a number of projects in the U.S. The San Bernardino County Medical Center in California is a five building isolated complex utilizing 400 high damping rubber bearings and 233 nonlinear viscous dampers with α = 0.5. Construction of the complex is scheduled to begin in 1994. Furthermore, studies for the seismic retrofit of the suspended part of the Golden Gate bridge in San Francisco concluded that the use of fluid dampers with α = 0.75 produce the desired performance (Rodriquez 1994).

The suitability of fluid viscous dampers for enhancing the seismic resistance of structures has been studied by Constantinou (1992, 1993). Fluid dampers with an orifice coefficient α=1 were tested over the temperature range 32°F to 122°F (0°C to 50°C). The dampers tested exhibited variations of their damping constant from a certain value at room temperature (75°F, 24°C) to + 44% of that value at 32°F (0°C) to -25% of that value at 122°F (50°C). This rather small change in properties over a wide range of temperature is in sharp contrast to the extreme temperature dependency of viscoelastic solid dampers. Figure 10 shows experimental results on the mechanical properties of a tested fluid damper over the temperature range of 0° to 50°C ($P = C_o \dot{u}$).

The inclusion of fluid viscous dampers in the tested structures on a shake table resulted in reductions in story drifts of 30% to 70%. These reductions are comparable to those achieved by other energy dissipating systems such as viscoelastic, friction and yielding steel dampers. However, the use of fluid dampers also resulted in reductions of story shear forces by 40% to 70% while other energy absorbing devices were incapable of achieving any comparable reduction. The reason for this difference is the nearly pure viscous behavior of the fluid dampers tested.

Figure 11 shows the construction of the nonlinear viscous dampers of the San Bernardino County Medical Center. These dampers have output force of 1400 kN at velocity of 1500 mm/s and a stroke of ± 610 mm. Scaled versions of these damper (scale of 1/6) have been very recently tested at the University at Buffalo. Figure 12 shows the recorded force-velocity data for one of the tested dampers. The nonlinear characteristics of the damper are evident. It may be noted in this figure that the output of the device is nearly unaffected by temperature in the range of 0 to 50° C. This temperature insensitivity has been a specific project requirement. It has been achieved entirely by passive means, that is by compensation of the changes in fluid properties with changes in the volume of the fluid and metallic parts.

DISSIPATION OF ENERGY IN SYSTEMS WITH NONLINEAR VISCOUS DAMPERS

Consider a fluid viscous damper with constitutive relation described by Equation (14). The energy dissipated in a cycle of sinusoidal motion

$$u = u_o \sin \omega_o t \tag{15}$$

is

$$W_d = \int_o^T P \dot{u} \, dt \tag{16}$$

where $T = 2\pi / \omega_o$, u_o = amplitude and ω_o = frequency. Integration of Equations (14) to (16) results in

$$W_d = 4 \cdot 2^\alpha \cdot \frac{\Gamma^2 \left(1 + \frac{\alpha}{2}\right)}{\Gamma(2 + \alpha)} \cdot C \cdot u_o^{1+\alpha} \cdot \omega_o^\alpha \tag{17}$$

where Γ is the gamma function. The dissipated energy may also be expressed in terms of the peak value of damper force $P_{max} = C u_o^\alpha \omega_o^\alpha$. That is,

$$W_d = 4 \cdot 2^\alpha \cdot \frac{\Gamma^2 \left(1 + \frac{\alpha}{2}\right)}{\Gamma(2 + \alpha)} \cdot P_{max} u_o \tag{18}$$

Equation (18) may be used to demonstrate the advantages of nonlinear viscous dampers with small values of parameter α. Consider two cases, one with $\alpha = 0.5$ (a technologically advanced design) and one with $\alpha = 2$ (a typical design with cylindrical or Bernoulian orifices). For the case of $\alpha = 0.5$, $W_d = 3.496 \, P_{max} u_o$, whereas for the case of $\alpha = 2$, $W_d = 2.667\, P_{max}u_o$. Thus for the same level of output force and amplitude of motion, the $\alpha = 0.5$ damper dissipates 31% more energy than the $\alpha = 2$ damper.

The significance of this difference in energy dissipation capability is more conveniently demonstrated by studying the behavior of a single-degree-of-freedom system with dampers. Let the mass of the system be M and the stiffness (linear and elastic) be K. A damper with characteristics described by Equation (14) is included in the system. The damping ratio is defined by

$$\xi = \frac{W_d}{2\pi K u_o^2} \tag{19}$$

in which u_o = amplitude of harmonic motion at the undamped natural frequency $\omega_o = \sqrt{K/M}$. By virtue of Equation (17), the damping ratio is

$$\xi = \frac{2^{1+\alpha} C u_o^{\alpha-1} \omega_o^{\alpha-2}}{\pi M} \cdot \frac{\Gamma^2\left(1 + \frac{\alpha}{2}\right)}{\Gamma(2 + \alpha)} \tag{20}$$

Specifically, for $\alpha = 0.5$

$$\xi = 0.55641 \frac{C}{M u_o^{1/2} \omega_o^{3/2}} \tag{21}$$

for $\alpha = 1$

$$\xi = \frac{C}{2M\omega_o} \tag{22}$$

and for $\alpha = 2$

$$\xi = \frac{4 C u_o}{2\pi M} \tag{23}$$

As expected the damping ratio is dependent on the amplitude of motion. For dampers with $\alpha < 1$, the damping ratio reduces with increasing amplitude of motion. The opposite is true for dampers with $\alpha > 1$, whereas for linear dampers ($\alpha = 1$) the damping ratio is independent of amplitude of motion. To illustrate this behavior, Figure 13 shows the damping ratio of a single-degree-of-freedom system with weight

W = 7000 kN (M = 713557 Kg) and undamped frequency ω_o = $\pi\, rad/s$. Three different cases of dampers are considered, each of which produces a peak force equal to 704 kN (or 0.1 W) at velocity of 785 mm/s. That is, damping constant C is equal to 25.13 kN $s^{1/2}/mm^{1/2}$ for α = 1/2, 0.8967 kN s/mm for α = 1 and 0.001142 kN s^2/mm^2 for α = 2.

Figure 13 depicts the damping ratio of the system with α = 0.5 approaching infinity as the amplitude of motion tends to zero. This is not realistic. Actual devices exhibit linear behavior at low levels of velocity so that the damping ratio levels off to a constant value, as shown in the figure.

The behavior of the three systems under free vibration and seismic excitation is shown in Figures 14 and 15, respectively. Figure 14 compares the time histories of displacement of the three systems when they are released from a 250 mm initial displacement. The differences in energy dissipation capability among the three systems are evident in the reduction of amplitude per cycle of motion. Figure 15 compares force-displacement loops (force is total force, that is the restoring plus damping force) of the three systems when excited by the SOOE component of the 1940 El Centro earthquake scaled to a peak ground acceleration of 0.68 g. The better performance of the nonlinear α = 0.5 damper is apparent. Comparing the α = 0.5 damper system to the linear viscous system (α = 1) we observe a nearly 20% reduction in displacement and a 10% reduction in total force. This better performance of the α = 0.5 system is actually achieved with a peak damper force lower than that of the other systems. Specifically, for α = 0.5 the maximum damper force is 640 kN, for α = 1 the force is 700.5 kN and for α = 2 the force is 930.5 kN.

CONSIDERATIONS IN THE DESIGN OF ENERGY ABSORBING SYSTEMS

The presented review of energy absorbing systems demonstrates that these systems are capable of producing significant reduction of interstory drift in the moment-resisting frames in which they are installed. Accordingly, they are all suitable for seismic retrofit applications in existing buildings.

Let us consider the implications of the use of energy absorbing systems in an existing moment-resting frame building. The gravity-load-carrying elements of the structural system have sufficient stiffness and strength to carry the gravity loads and, say, seismic forces in a moderate earthquake. The energy absorbing devices are installed in new bracing systems and, say, are capable of reducing drifts to half of those of the original system in a severe earthquake. An immediate observation is that the reduction of drift will result in a proportional reduction in bending moment in the columns,provided that the behavior is elastic, which will now undergo limited rather than excessive yielding.

However, the behavior of the retrofitted structure has changed from that of a moment-resisting frame to that of a braced frame. The forces which develop in the energy absorbing elements will induce additional axial forces in the columns. Depending on the type of energy absorbing device used, this additional axial force may be in-phase with the peak drift and, thus, may affect the safety of the loaded column.

Figure 16 shows idealized force-displacement loops of various energy absorbing devices. In the friction and steel yielding devices, the peak brace force occurs at the time of peak displacement. Accordingly, the additional column force, which is equal to $F\sin\theta$ (θ is the brace angle with respect to the horizontal), is in-phase with the bending moment due to column drift. Similarly, in the viscoelastic device a major portion of the additional column force is in-phase with the bending moment. In contrast, in the viscous device the additional column force is out-of-phase with the bending moment.

The implications of this difference in behavior of energy absorbing devises are illustrated in Figure 17. We assume that the energy absorbing devices are installed in the interior columns of a reinforced concrete frame. The nominal axial force-bending moment interaction diagram of a column is shown. It is assumed that the column was designed to be in the compression controlled range of the diagram. During seismic excitation, the moment-resisting frame undergoes large drifts and column bending moments but axial load remains practically unchanged. Failure will occur when the tip of the P-M loop reaches the nominal curve as illustrated in Figure 17a. The available capacity of the column is related to the distance between the tip of the P-M loop and the nominal curve (shown as a dashed line in Figure 17).

In the frame with added energy dissipating devices, the P-M loops show less bending moment. Despite this, the available capacity of the column may not have increased since the distance between the tip of the P-M loop and the nominal curve may have remained about the same. An exception to this behavior can be found in the viscous device.

The conclusion of the preceding discussion is that drift is not the only concern in design. Energy absorbing devices may reduce drift and thus reduce inelastic action. However, depending on their force-displacement characteristics, they may induce significant axial column forces which may lead to significant column compression or even column tension. This concern is particularly important in the seismic retrofitting of structures which suffered damage in previous earthquakes. After all, it may not always be possible to upgrade the seismic resistance of such structures by the addition of energy absorbing devices alone. It may also be necessary to strengthen the columns.

Constantinou (1992) utilized the experimental results of Aiken (1993) to demonstrate that the addition of friction dampers (Sumitomo type) to a tested 9-story structure resulted in significant additional axial load to the interior columns. Specifically, this additional axial load was 130% of the axial load due to gravity (gravity load = 12.8 kips, total axial load = 29.5 kips). Furthermore, similar calculations for a 3-story structure with ADAS elements tested by Whittaker (1989), resulted in additional axial load of only 14% of the gravity load.

CONCLUSIONS

Supplemental damping devices are capable of producing significant reductions of interstory drifts in the moment-resisting frames in which they are installed. They are suitable for applications of seismic retrofit of existing structures.

The behavior of structures retrofitted with supplemental damping devices changes from that of a moment-resisting frame to that of a braced frame. The forces which develop in the devices induce additional axial forces in the columns. For frictional, steel yielding and viscoelastic devices this additional axial force occurs in-phase with the peak drift and, thus, affects the safety of the loaded columns. This represents an important consideration in design and may impose limitations on the use of these devices in tall buildings. Exemption to this behavior can be found in a certain type of fluid damper which exhibits essentially viscous behavior.

REFERENCES

1.   Aiken, I.D. Kelly, J.M. (1988). "Experimental study of friction damping for steel frame structures." Proc. PVP Conference, ASME, Pittsburgh, PA, Vol. 133, 95-100.

2.   Aiken, I.D. and Nims, D.K., Whittaker, A.S., and Kelly, J.M. (1993), "Testing of passive energy dissipation systems," Earthquake Spectra, 9(3), 335-370.

3.   Arima, F., Miyazaki, M., Tanaka, H. and Yamazaki, Y. (1988). "A study on building with large damping using viscous damping walls." Proc., 9th World Conference on Earthquake Engineering, Tokyo-Kyoto, Japan, 5, 821-826.

4.   Ashour, S.A. and Hanson, R.D. (1987), Elastic Seismic Response of Buildings with Supplemental Damping, Report No. UMCE 87-01, The University of Michigan, Ann Arbor, MI.

5.   Bergman, D.M. and Hanson, R.D. (1993), "Viscoelastic mechanical damping devices tested at real earthquake displacements," Earthquake Spectra, 9(3), 389-418.

6.   British Standards Institution - BSI (1979). "Commentary on corrosion at bimetallic contacts and its alleviation." Standard PD 6484:1979, London, U.K.

7.   Caldwall, D.B. (1968), "Viscoelastic damping devices proving effective in tall buildings," AISC Engineering Journal, 23(4), 148-150.

8.   Chang, K.C., Lai, M.L., Soong, T.T., Hao, D.D. and Yeh, Y.C. (1993a), Seismic behavior and design guidelines for steel frame structures with added viscoelastic dampers, NCEER 93-0009, National Center for Earthquake Engineering Research, Buffalo, NY.

9.   Chang, K.C., Shen, K.L., Soong, T.T. and Lai, M.L. (1994), "Seismic retrofit of a concrete frame with added viscoelastic dampers," 5th National Conference on Earthquake Engineering, Chicago, IL.

10.   Chang, K.C. Soong, T.T., Lai, M.L., Neilsen, E.J. (1993b), "Viscoelastic dampers as energy dissipation devices for seismic applications," Earthquake Spectra, 9(3), 371-388.

11.   Chang, K.C., Soong, T.T., OH, S-T. and Lai, M.L. (1992), "Effect of ambient temperature on a viscoelastically damped structure," ASCE Journal of Structural Engineering, 118(7), 1955-1973.

12.   Constantinou, M.C., Reinhorn, A.M., Mokha, A. and Watson, R. (1991a). "Displacement control device for base-isolated bridges." Earthquake Spectra, 7(2), 179-200.

13.   Constantinou, M.C., Kartoum, A., Reinhorn, A.M. and Bradford, P. (1991b). "Experimental and theoretical study of a sliding isolation system for bridges." Report No. NCEER 91-0027, National Center for Earthquake Engineering Research, Buffalo, NY.

14.   Constantinou, M.C. and Symans, M.D. (1992). "Experimental and analytical investigation of seismic response of structures with supplemental fluid viscous dampers. "Report No. NCEER-92-0032, National Center for Earthquake Engineering Research, Buffalo, NY.

15. Constantinou, M.C., Symans, M.D., Tsopelas, P. and Taylor, D.P. (1993). "Fluid viscous dampers in applications of seismic energy dissipation and seismic isolation." Proc. ATC-17-1 Seminar on Seismic Isolation, Passive Energy Dissipation, and Active Control, San Francisco, March.

16. Ferry, J.D. (1980), Viscoelastic Properties of Polymers, John Wiley, New York, NY.

17. Filiatrault, A. and Cherry, S. (1985). "Performance evaluation of friction damped braced steel frames under simulated earthquake loads." Report of Earthquake Engineering Research Laboratory, University of British Columbia, Vancouver, Canada.

18. Fitzgerald, T.F., Anagnos, T., Goodson, M. and Zsutty, T., (1989). "Slotted bolted connections in aseismic design of concentrically braced connections. " Earthquake Spectra, 5(2), 383-391.

19. Foutch, D.A., Wood, S.L. and Brady, P.A. (1993), "Seismic retrofit of nonductile reinforced concrete frames using viscoelastic dampers," ATC-17-1 on Seismic Isolation, Passive Energy and Active Control, 2, 605-616.

20. Fujita, T. (editor) 1991). "Seismic isolation and response control for nuclear and non-nuclear structures." Special Issue for the Exhibition of the 11th International Conference on Structural Mechanics in Reactor Technology, SMiRT 11, Tokyo, Japan.

21. Fujita, S. Fujita, T. Furuya, O., Morikawa, S., Suizu, Y., Teramoto, T. and Kitamura, T. (1992), "Development of high damping rubber damper for vibration attenuation of high-rise buildings," Proc. 10th World Conf. Earthquake Engrg., 2097-2101, Balkema, Rotterdam.

22. Grigorian, C.E. and Popov, E.P. (1993). "Slotted bolted connections for energy dissipation." Proc. ATC-17-1 Seminar on Seismic Isolation, Passive Energy Dissipation, and Active Control, San Francisco, March.

23. Hsu, S-Y. and Fafitis, A., (1992). "Seismic analysis and design of frames with viscoelastic connections." J. Struct. Engrg., ASCE, 118(9), 2459-2474.

24. Kasai, K., Munshi, J.A., Lai, M.L. and Maison, B.F. (1993), "Viscoelastic damper hysteretic model: Theory, experiment and application," Proc. ATC-17-1 on Seismic Isolation, Energy Dissipation, and Active Control, 2, 521-532.

25.   Kelly, J.M., Skinner, M.S. and Beucke, K.E., (1980). "Experimental testing of an energy-absorbing base isolation system. " Report No. UCB/EERC-80/35, University of California, Berkeley.

26.   Kelly, J.M. (1988). "Base isolation in Japan, 1988". Report No. UCB/EERC-88/20, University of California, Berkeley.

27.   Kirekawa, A., Ito, Y. and Asano, K. (1992), "A study of structural control using viscoelastic material," Proc. 10th World Conf. Earthquake Engrg., 2047-2054, Balkema, Rotterdam.

28.   Lin, R.C., Liang, Z., Soong, T.T. and Zhang, R.H. (1991), "An experimental study of seismic structural response with added viscoelastic dampers," Engineering Structures, 13, 75-84.

29.   Lobo, R.F., Bracci, J.M., Shen, K.L., Reinhorn, A.M. and Soong, T.T. (1993), "Inelastic response of R/C structures with viscoelastic braces," Earthquake Spectra, 9(3), 419-446.

30.   Makris, N. and Constantinou, M.C. (1992). "Spring-viscous damper systems for combined seismic and vibration isolation." Earthquake Engrg. Struct. Dyn., 21 (8), 649-664.

31.   Pall, A.S. and March, C. (1982). "Response of friction damped braced frames." J. Struct. Engrg., ASCE, 108(60, 1213-1323.

32.   Pall, A.S. , Verganelakis, V. and March, C. (1987). "Friction dampers for seismic control of concordia university library building. "Proc. 5th Canadian Conference on Earthquake Engineering, Ottawa, Canada, 191-200.

33.   Pall, A.S. and Pall, R. (1993). "Friction dampers used for seismic control of new and existing buildings in Canada." Proc. ATC-17-1 Seminar on Seismic Isolation, Passive Energy Dissipation, and Active Control, San Francisco, March.

34.   Rodriguez, S., Seim, C. and Ingham, T. (1994). "Earthquake protective systems for the seismic upgrade of the Golden Gate bridge." Proc. 3rd U.S.-Japan Workshop on Protective Systems for Bridges," Berkeley, CA, January.

35.   Roeder, C.W. and Popov,, E.P. (1978). "Eccentrically braced steel frames for earthquakes." J. Struct. Div., ASCE, 104(3), 391-412.

36.   Shen, K.L. and Soong, T.T. (1994), "Modeling of viscoelastic dampers for structural applications," ASCE J. Eng. Mech., submitted.

37.   Skinner, R.I., Tyler, R.G., Heine, A.J., and Robinson, W.H. (1980). "Hysteretic dampers for the protection of structures from earthquakes." Bulletin of New Zealand National Society for Earthquake Engineering, 13(1), 22-36.

38.   Su, Y-F. and Hanson, D. (1990), Seismic Response of Building Structures with Mechanical Damping Devices, Report No. UMCE 90-02, University of Michigan, Ann Arbor, MI.

39.   Tsai, C.S. and Lee, H.H. (1993), "Application of viscoelastic dampers to high-rise buildings," ASCE J. Struct. Eng., 119(4), 1222-1233.

40.   Tyler, R.G. (1978). "Tapered steel energy dissipators for earthquake resistant Structures." Bulletin of New Zealand National Society for Earthquake Engineering, 11(4), 282-294.

41.   Tyler, R.G. (1985). "Further notes on a steel energy-absorbing element for braced frameworks." Bulletin of New Zealand National Society for Earthquake Engineering, 18(3), 270-279.

42.   Uang, C-M. and Bertero, V.V. (1988). "Use of energy as a design criterion in earthquake-resistant design." Report No. UCB/EERC-88/18, University of California, Berkeley.

43.   Whittaker, A.S., Bertero, V.V., Alonso, J.L. and Thompson, C.L. (1989). "Earthquake simulator testing of steel plate added damping and stiffness elements." Report No. UCB/EERC-89/02, University of California, Berkeley.

44.   Zhang, R.H., Soong, T.T. (1992), "Seismic design of viscoelastic dampers for structural applications," ASCE Journal of Structural Engineering, 118(5), 1375-1392.

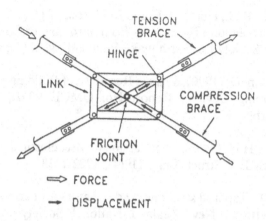

Figure 1   Friction Damper of Pall (1982).

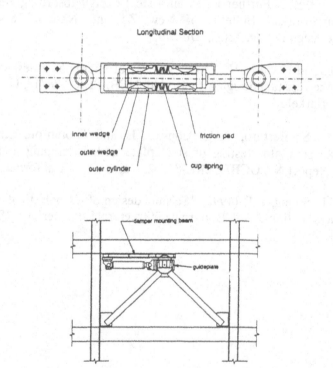

Figure 2   Sumitomo Friction Damper and Installation Detail (from Aiken
                  1990).

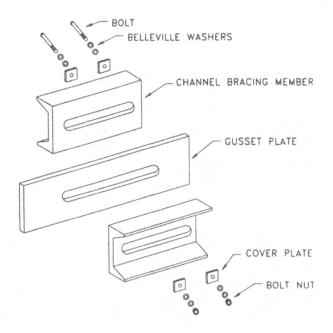

Figure 3   Slotted Bolted Connection of Fitzgerald (1989).

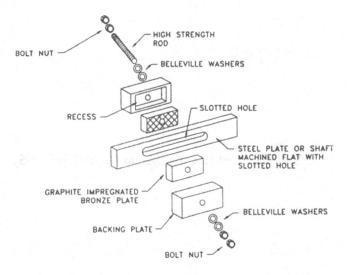

Figure 4   Friction Assembly in Displacement Control Device of Constantinou
                 (1991a).

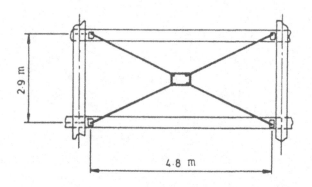

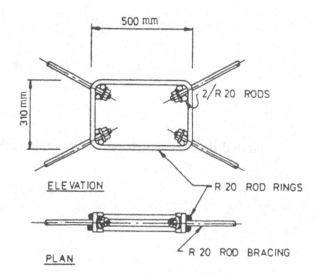

Figure 5   Yielding Steel Bracing System (from Tyler 1985).

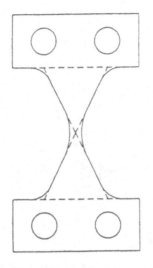

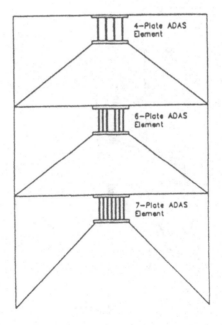

Figure 6   ADAS Element and Installation Detail (from Whittaker 1989).

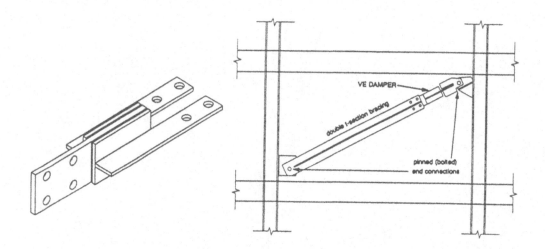

Figure 7   Viscoelastic Damper and Installation Detail (from Aiken 1990).

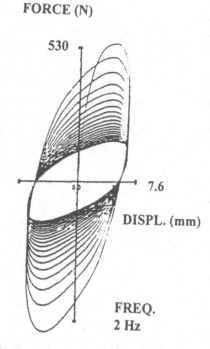

Figure 8   Force-displacements Loops of Viscoelastic Damper under Cyclic
                  Loading (from Kasai 1993).

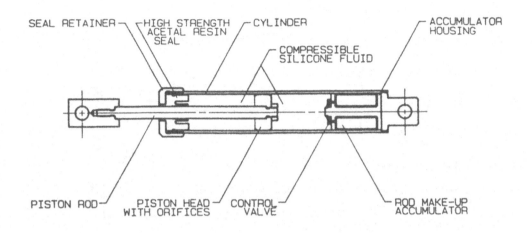

Figure 9   Construction of Fluid Viscous Damper with Accumulator.

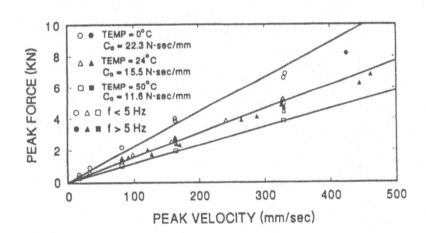

Figure 10   Effect of Temperature on Mechanical Properties of Linear Fluid Viscous Damper.

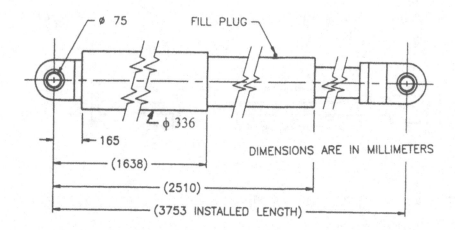

Figure 11   Construction of Nonlinear Fluid Viscous Damper for the San Bernardino County Medical Center, CA.

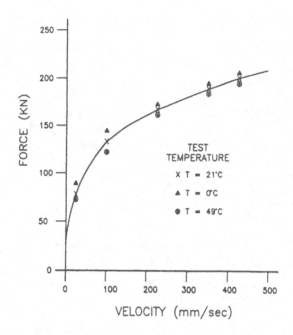

Figure 12   Force-Velocity Relation of 1/6 Scale Prototype Fluid Damper of San Bernardino County Medical Center.

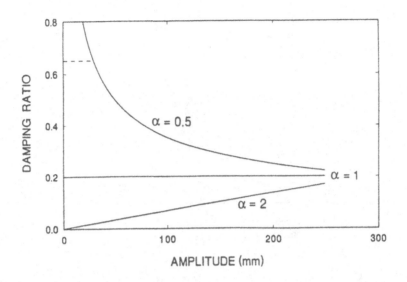

Figure 13   Damping Ratio of Systems with Nonlinear Fluid Dampers as Function of Amplitude of Motion.

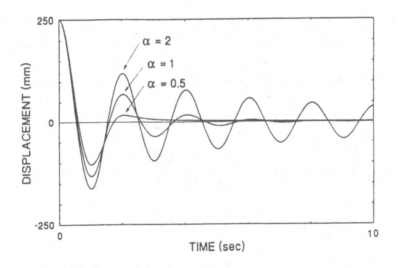

Figure 14   Time Histories of Motion of Systems with Nonlinear Fluid Dampers when Released from an Initial Displacement.

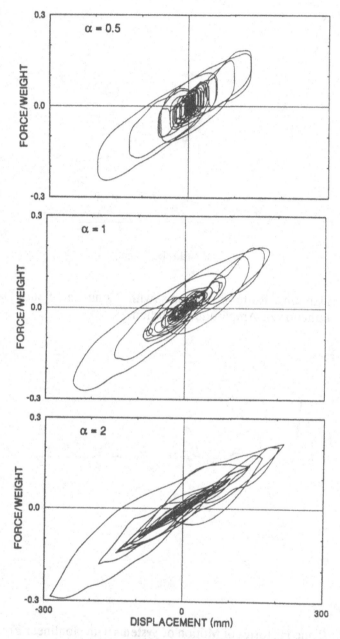

Figure 15   Force-displacement Loops of Response of Systems with Nonlinear
                   Viscous Dampers when Subjected to the 1940 El Centro,
                   Component S00E Earthquake, Scaled to 0.68g.

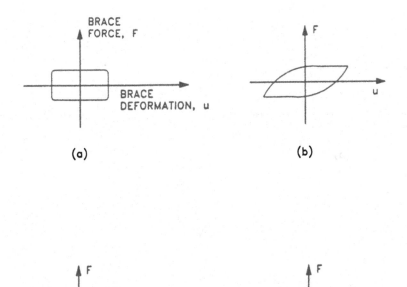

Figure 16   Force-displacement Loops of (a) Friction Device, (b) Steel Yielding Device, (c) Viscoelastic Device, (d) Viscous Device.

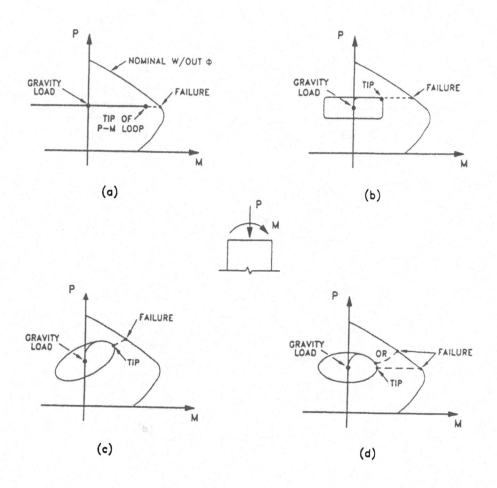

Figure 17   Column Interaction Diagrams and Axial Force- Bending Moment
                   Loops during Seismic Excitation for (a) Moment-resisting Frame,
                   (b) Friction Damped Frame, (c) Visco-elastically Damped Frame,
                   and (d) Viscously Damped Frame.

CHAPTER XI

PRINCIPLES OF TMD AND TLD
- BASIC PRINCIPLES AND DESIGN PROCEDURE -

H. Iemura
Kyoto University, Kyoto, Japan

ABSTRACT

In this chapter, the principles of TMD (Tuned Mass Damper) are explained. Selection of structural parameters of TMD is discussed based on Den Hartog's optimum theory for a simplified 2-DOF (degree-of-freedom) system. Design procedure of TMD for implementation to structures is also introduced. Discussions are extended to TLD (Tuned Liquid Damper). Shaking-table tests of TLD for flexible structures at Kyoto University are also introduced.

Principle of TMD (Tuned Mass Damper)

With recent development in computer-based structural design and high-strength materials, structures are becoming more flexible and lightly damped. When subjected to dynamic loads such as traffic load, wind, earthquake, wave, vibration lasting for long duration may be easily induced in this type of structures. To increase comfort of working people, function of installed machineries and equipments, and reliability of structures, damping capacity of structures in the elastic region should be increased.

If we have a fixed reaction wall adjacent to the top of a structure as shown in Fig. 1, viscous or frictional damper can be installed effectively to increase damping capacity of the structure. However, this is usually impossible because flexible structures are very tall and no fixed point is available.

TMD is a vibration system with mass m_T, spring K_T and viscous c_T usually installed on the top of structures as shown in Fig. 2. When the structure starts to vibrate, TMD is excited by the movement of the structure. Hence, kinetic energy of the structure goes into TMD system to be absorbed by the viscous damper of TMD. To achieve the most efficient energy absorbing capacity of TMD, natural period of TMD by itself is tuned with the natural period of the structure by itself, from which the system is called "Tuned Mass Damper". The viscous damper of TMD shall also be adjusted to the optimum value to maximize the absorbed energy. TMD is a mechanically simple system which does not need any external energy supply for operation. Because of easy maintenance and high reliability, TMD is used in many flexible and lightly-damped towers, buildings and so on in Japan.

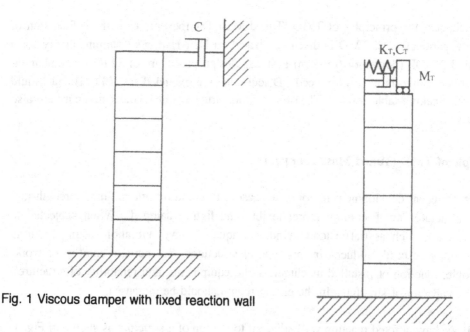

Fig. 1 Viscous damper with fixed reaction wall

Fig. 2 Tuned mass damper on a structure

Principles of TMD and TLD 243

Determination of the Optimum TMD

One TMD is effective in reducing dynamic response of only a single vibration mode of the structure. Although a structure has many vibration modes in reality, basic properties of TMD can be clearly discussed using a simplified 2-DOF model consisting of the main structure and the TMD system (Fig. 3).

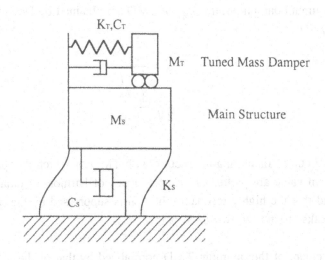

Fig. 3 2-DOF modelling of main structure and tuned mass damper system

Let us define the following parameters to be used in the following discussion.

Natural Frequency of TMD $\omega_T = \sqrt{\dfrac{K_T}{m_T}}$

Damping Ratio of TMD $\xi_T = \dfrac{c_T}{2m_T\omega_T}$

Natural Frequency of Main Structure $\omega_s = \sqrt{\dfrac{K_s}{m_s}}$

Damping Ratio of Main Structure $\xi_s = \dfrac{c_s}{2m_s\omega_s}$

Mass Ratio $\mu = m_T/m_s$

Frequency (Tuning) Ratio $\gamma = \omega_T/\omega_s$

When $\xi_T = 0$, a 2-DOF system shows 2 uncoupled vibration modes and, when $\xi_T = \infty$, the 2-DOF system becomes a 1-DOF vibration system. Steady-state dynamic response subjected to harmonic excitation can be obtained analytically. It is usually called the resonant curve or dynamic magnification factor (DMF) curve plotted against angular frequency of the harmonic excitation. It is interesting to notice that DMF curves cross two fixed points independent of the damping ratio ξ_T.

Den Hartog defined the optimum TMD by letting the two fixed points the same value and as high as possible in the DMF curve. The physical meaning of this is to obtain flat DMF curve at the resonant frequency, and consequently to suppress the dynamic response of the main structure most effectively. From this definition, the optimum frequency ratio γ_{opt} and the optimum damping ratio $\xi_{T_{opt}}$ of TMD are obtained by Den Hartog as function of mass ratio μ, i.e.,

$$\gamma_{opt} = \frac{1}{1+\mu} \tag{1}$$

$$\xi_{T_{opt}} = \frac{1}{2}\sqrt{\frac{3\mu/2}{1+3\mu}} \tag{2}$$

The DMF curves of the main structure with TMD of which damping ratio $\xi_T = 0$, ∞, and optimum value are plotted against frequency of harmonic excitation in Fig. 4. It is clearly found that the highly resonant vibration is suppressed by the optimum TMD to the low two peaks around the two fixed points.

The movement of the optimum TMD normalized by that of the main structure corresponding to Fig. 4 is plotted in Fig. 5. At the resonant frequency of the optimum TMD, the movement of the TMD is 8 times larger than that of the main structure, from which

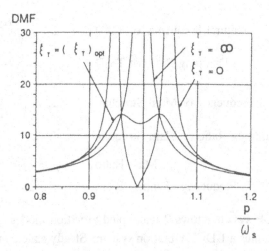

Fig. 4   Resonance curve of the structure with TMD

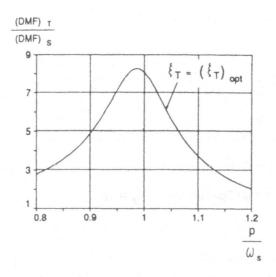

Fig. 5   Normalized resonance curve of TMD

the TMD is absorbing the kinetic energy of the main structure most effectively. However, for actual implementation, a large space is needed in order to accommodate the movement of the TMD.

With the optimum TMD, the damping ratio of the main structure will be increased by the amount of

$$\Delta \xi_{eq} = \frac{1}{2} \sqrt{\frac{\mu/2}{1 + \mu/2}} \qquad (3)$$

Figs. 6(a), (b), (c) show γ_{opt}, $\xi_{T_{opt}}$, and $\Delta \xi_{eq}$ respectively against mass ratio μ. Curves (a) in each figure show the cases for the harmonic excitation. These figures can be used conveniently to approximately determine structural parameters of the TMD for implementation.

The above discussion on the optimum TMD is based on the steady-state response subjected to harmonic excitation. As reviewed by Yamaguchi et al (1991), the values of γ_{opt}, $\xi_{T_{opt}}$, and $\Delta \xi_{eq}$ are slightly different for free vibration, wind-induced self-oscillation, and forced random vibration, and these effects are shown by curves (b), (c) and (d). The values are also influenced by the damping ratio ξ_s of the main structure.

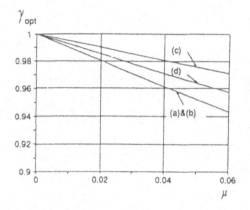

(a) Optimum Frequency Ratio

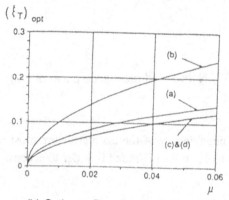

(b) Optimum Damping Ratio

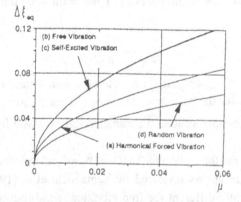

(c) Equivalent Damping Ratio

Fig. 6 Parameters of optimum TMD and equivalent structural damping
[Yamaguchi et al, 1991]

Design Procedure of TMD

In the design of an actual TMD, there are many design constraints arising from structural properties of the main structure. Following is a basic and fundamental design procedure of TMD for general flexible and lightly-damped structures.

1. Identify dynamic structural properties of a main structure.
   Determine natural frequency, vibration mass, damping ratio of the specific vibration mode of the structure to be controlled by TMD.

2. Determine design damping ratio $\Delta \xi_{eq}$ to be generated by TMD.

3. Assume appropriate mass ratio μ (determined mass m_T of TMD).

4. Calculate γ_{opt} and $\xi_{T_{opt}}$ from μ (determined stiffness K_T and damping coefficient C_T of TMD)

5. Calculate maximum displacement of TMD. If it is too large, go back to 3.

6. Design mechanical system of TMD including the methods on how to adjust ω_T and ξ_T.

7. Verification tests of TMD with shaking table or with other methods.

8. Installation of TMD to be main structure

9. Monitor behavior of TMD and adjust ω_T and ξ_T on site

The following issues shall be examined, if necessary:

- If the vibration mode of the structure to be controlled by TMD has another vibration mode close to it, 3- or higher-DOF modelling of the main structure and the TMD has to be used to determine the optimum TMD.

- Multiple TMD of which ω_T are distributed around the vibration period of the structure has robustness in tuning effect.

- If the displacement of TMD is too large, use of larger mass or higher damping becomes inevitable, which may not always give the optimum TMD.

Types of TMD

There are many types of TMD for implementation. The following (Fig. 7) are some examples. Innovative challenge is highly expected in this field.

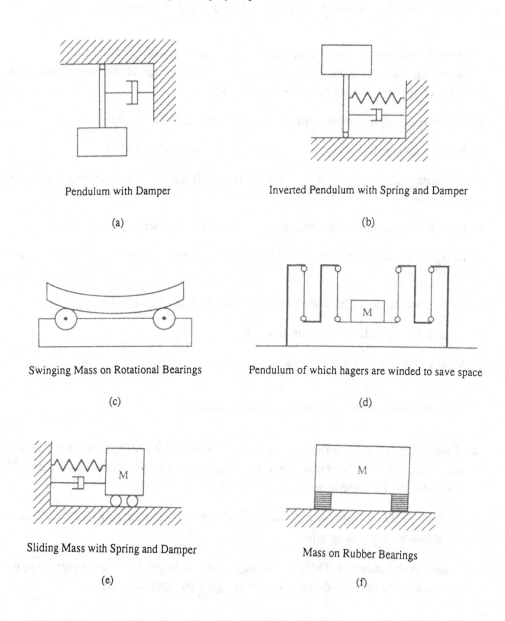

Pendulum with Damper

(a)

Inverted Pendulum with Spring and Damper

(b)

Swinging Mass on Rotational Bearings

(c)

Pendulum of which hagers are winded to save space

(d)

Sliding Mass with Spring and Damper

(e)

Mass on Rubber Bearings

(f)

Fig. 7   Examples of different types of TMD

Principles and Properties of TLD (Tuned Liquid Damper)

Tuned liquid dampers which have been widely used in ships are recently implemented for vibration control of structures. A TLD uses water or other liquids as the moving mass and restoring force is generated by gravity. Energy absorption comes from boundaries between liquid and containers and turbulence in the liquid flow. The basic principle of TLD to absorb kinetic energy of the main structure is the same as TMD. Favorable properties of TLD compared to TMD are as follows:

- Because of no mechanical friction, smooth movement in small vibration is possible. Hence effective to control small movement.

- No complex mechanisms, thereby reasonable in cost and maintenance.

- Can be applied easily to two horizontal vibrations with a single TLD.

- Can be compact and portable, if large numbers are used.

On the other hand, an unfavorable property of TLD is relatively small mass of water compared to the mass of TMD which is usually made of steel. Hence, to have the same damping effect as TMD, larger space is required.

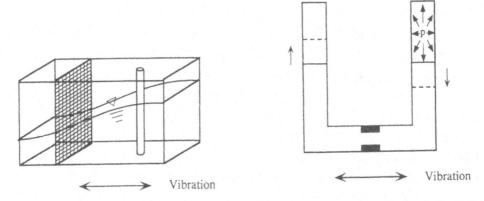

(a) Tuned Sloshing Damper with Meshes and Rods   (b) Tuned Liquid Column Damper with Orifice

Fig. 8   Types of TLD

Types of TLD

TLD can be divided into two categories. First is the sloshing damper as shown in Fig. 8(a). Vibration period is adjusted by the size of the container or depth of the liquid. Damping capacity is increased by placing meshes or rods in the liquid. The second is the tuned liquid column damper as shown in Fig, 8(b). Vibration period is adjusted by shapes of column or air pressure in the column. Damping capacity is increased by adjusting the orifice in the column which generates high turbulence.

Shaking Table Tests of TLD at Kyoto University

At the Dept. of Civil Engineering of Kyoto University, effects of a tuned sloshing damper to suppress earthquake response of a 3-story lightly-damped steel frame model are tested on a shaking table as shown in Fig. 9. Fundamental period and damping ratio of the structural model are 0.69 sec and 0.36% respectively. Sloshing of water in cylindrical vessel is used as TLD. The model is subjected to the 1940 El Centro NS record of which maximum acceleration is scaled to 25 cm/sec^2.

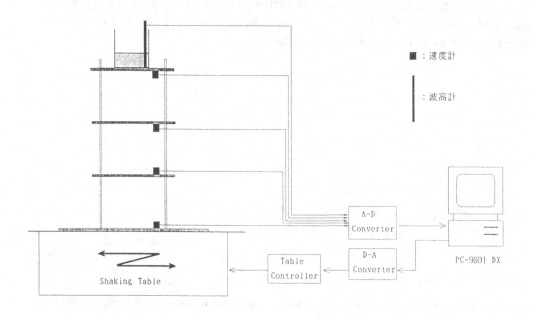

Fig. 9   Shaking-table tests of structural model with TMD

Figs. 10(a), (b) show velocity response of the 3rd floor of the model with and without TLD, and sloshing wave height time history. Over a few cycles at the beginning of oscillation, not much difference is found between the case with and without TLD. In these transient cycles, wave height grew larger, where TLD is less effective.

Fourier spectrum of velocity response shown in Fig. 10(a) is plotted in Fig. 11. Sharp peak of the first mode of the flexible structural model is suppressed and divided into two peaks due to the effect of the TLD. Relatively deep gap between the two peaks is probably due to small damping in the TLD. The second mode response is not affected by the TLD at all.

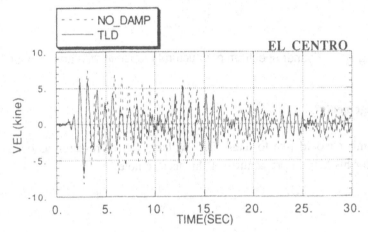

(a) velocity response of the structural model with and without TLD

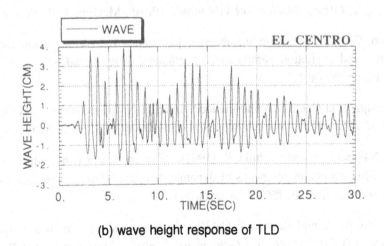

(b) wave height response of TLD

Fig. 10   Earthquake response under El Centro earthquake record

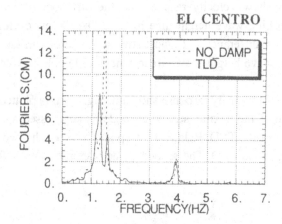

Fig. 11   Fourier spectrum of structural response with and without TLD

Acknowledgements

The author would like to thank Dr. William Tanzo, research associate at Saitama University for his help in preparation of this lecture note.

References

Den Hartog, J. (1956). *Mechanical Vibrations.* 4th ed., McGraw-Hill, New York.

Warburton, G. (1982). "Optimum absorber parameters for various combinations of response and excitation parameters," *Earthquake Engg. and Structural Dynamics,* vol.10, pp.381–401.

Tamura, Fujii, et al (1988). "Wind-induced vibration of tall towers and practical application of TSD," *Proc., Workshop on Serviceability of Buildings,* Ottawa.

Ueda, T., Nagasaki, R, and Koshida, K. (1991). "Suppression of wind-induced vibration in tower-like structures by means of dynamic dampers," *Proc., Colloquium on Control of Structures,* Part B, JSCE. (in Japanese)

Sakai, F., Takeda, S. and Tamaki, T. (1991). "Experimental study on tuned liquid column damper (TLCD)," *Proc., Colloquium on Control of Structures,* Part B, JSCE. (in Japanese)

Fujino, Y., Pacheco, B., Chaiseri, P., Sun, L.M. and Koga, K. (1990). "Understanding of TLD properties based on TMD analogy," *Journal of Structural Engineering*, JSCE, 36A.

Yamaguchi, H., Fujino, Y. and Tsumura, N. (1991). "Passive control of structures with TMD," *Proc., Colloquium on Control of Structures*, Part B, JSCE. (in Japanese)

Iemura, H., et al (1993). "Shaking-table tests on sloshing dampers for flexible structures," *Annual Convention of JSCE*. (in Japanese)

Iemura, H., et al (1992). "Comparison of passive, active and hybrid control techniques on earthquake response of flexural structures," *U.S.-Italy-Japan Workshop on Structural Control and Intelligent Structures*.

CHAPTER XII

PASSIVE ENERGY DISSIPATION DEVELOPMENT IN U.S.

M.C. Constantinou
State University of New York at Buffalo, Buffalo, NY, USA

ABSTRACT

Passive energy dissipation systems were developed in the United Stated either specifically for civil engineering applications or they evolved from devices and materials used in industrial, automotive, military and aerospace applications. Yielding steel and frictional devices were specifically developed for structural applications, whereas viscoelastic dampers and fluid viscous devices were developed for other applications and adapted for structural applications.

FRICTION DAMPERS

The earliest work on the development of friction dampers for reducing seismic motions in buildings appears to have been that of Keightley (1977,1979) at Montana State University. The friction dampers of Keightley consisted of steel plates clamped together with bolts and belleville washers. Various forms of lubricants were used to prevent locking of the interfaces. Of these only powdered oil shale was found to be satisfactory. Keightley's work was concluded with a number of unanswered questions and suggestions for future research. The questions on the possible relaxation of the clamping assembly, the long-term stability of the lubricant and corrosion of the steel to steel sliding interface could have prevented the use of this device today.

In the late 1970's the firm of Severud, Perrone, Sturn and Bendel of New York has installed two large friction dampers between the existing structure of the Gorgas hospital in the Panama Canal Zone and two exterior massive concrete pylons.

Experimental studies of slotted bolted connections started at San Jose State University and proceeded approximately concurrently with the research of Keightley. Fitzgerald (1989) reported on the results of this study. This slotted bolted connection consisted of a gusset plate, two back-to-back channels sections, cover plates and bolts with belleville washers as shown in Figure 1. The figure shows also typical force-displacement loops obtained in tests of the device. Recently, Grigorian (1993) conducted tests of a similar slotted bolted connection which consisted of a brass to steel sliding interface. Figure 2 shows a schematic of this device together with a representative force-displacement loop.

Constantinou (1991) described another friction device of which the interface consisted of graphite impregnated bronze in contact with stainless steel. Figure 3 shows the device and representative force-displacement loops. One should note the exceptional stability of the properties of the device over 200 cycles of testing (total travel = 20.3 m at peak velocity of 25 mm/s). This was the result of proper selection of the materials forming the interface and the use of graphite as solid lubricant.

Fluor Daniel developed a frictional device called Energy Dissipating Restraint (EDR). Shown in Figure 4, this device is characterized by self-centering capability as demonstrated in the loops of Figure 4 (Nims 1993).

The described devices utilize sliding interfaces consisting of steel on steel, brass on steel and graphite impregnated bronze on stainless steel. The composition of the interface is of paramount importance for ensuring the longevity of the device. Carbon and low allow steels (common steel) will corrode and thus the frictional properties of steel to steel interfaces will change with time. However, brass and bronze when in contact with common steel promote severe additional corrosion of steel and such interfaces should be avoided. Only stainless steels with high content of chromium do not suffer additional corrosion when in contact with brass or bronze (BSI 1979).

VISCOELASTIC DEVICES

Viscoelastic materials (acrylic polymers) have been originally developed for surface damping treatments (Nashif 1985). Devices made of bonded viscoelastic layers were developed by the 3 M Company and used in wind vibration control applications. The devices have been extensively studied for applications of seismic energy dissipation (Lin 1991, Aiken, 1993, Chang 1993, Lobo 1993).

Viscoelastic dampers exhibit viscoelastic solid behavior with strong dependencies on temperature, frequency and amplitude of motion. Particularly, the temperature dependency of these devices is an important design consideration and needs to be explicitly modeled for the dynamic analysis of viscoelastically damped structures. Kasai (1993) presented a fractional derivative formulation with temperature effects of the constitutive relation of the 3M viscoelastic material. The formulation is capable of capturing the viscoelastic and temperature effects with very good accuracy. Figure 5 demonstrates the accuracy of the model.

The 3M viscoelastic devices are for use as energy dissipating braces (see lecture notes on "Principles of Friction, Viscoelastic, Yielding Steel and Fluid Viscous Dampers: Properties and Design". A different approach to the use of viscoelastic materials has been proposed by the Lorant Group in Phoenix, Arizona and studied by Hsu (1992). Figure 6 shows a detail of a beam to column connection which incorporates viscoelastic materials. It has been used at a 2-story building in Phoenix, Arizona.

METALLIC DEVICES

A number of yielding steel devices has been tested at U.C. Berkeley in the late 1970's for use in seismic isolation systems and as damping supports for piping systems (Kelly 1980). While none of these devices was used in seismic isolation systems in the U.S., modifications of these devices have been recently studied for applications of energy dissipation. The ADAS devices are X-shaped double-clamped steel devices which have been tested as energy dissipation devices in buildings (Bergman 1987, Whittaker 1989). Figure 7 shows a hysteretic loop of a tested ADAS device (Whittaker 1989). The device exhibits stable hysteretic behavior for a large number of cycles. Repeated yielding of the device leads to fatigue failure.

Shape memory alloys have been studied at the University of California and the University at Buffalo (Whitting 1992). Materials such as Nitinol (nickel-titanium) and Cu-Zn-Al (copper-zinc-aluminum) undergo reversible phase transformation when deformed, so that they exhibit hysteretic behavior without yielding. The behavior of these materials is very desirable (hysteresis without yielding), however they are currently prohibitively expensive.

FLUID VISCOUS DAMPERS

Fluid dampers which operate by fluid flow through orifices were invented by Ralph Peo of Buffalo, N.Y. in 1925 and used in automotive applications. Taylor Devices of New York have been manufacturing such devices since 1955 with over two million units produced since then. These devices have been used in shock and vibration isolation systems of military and aerospace hardware, as wind dampers, as crane buffers and recently have been studied as energy dissipating devices in buildings and bridges (Constantinou 1992, 1993, Tsopelas 1994).

A variety of designs of fluid dampers has been developed in the United States. Figure 8 illustrates the four basic design characteristics. The fluidic device uses specially shaped orifices to achieve a force output

$$F = C|\dot{u}|^{\alpha} sgn(\dot{u}) \tag{1}$$

where F = force, C = damping constant, $\dot{u}$ = velocity and α = coefficient in the range of 0.5 to 2. The value α = 2 is achieved with cylindrical orifices, a performance which is typically unacceptable. A value α = 0.5 is particularly effective in attenuating high velocity pulses, as those expected in near fault earthquake excitations. A value of α = 1 results in linear viscous behavior which is usually desirable in applications of seismic energy dissipation. Figure 9 shows recorded force-displacement loops of a device with α = 0.5 from tests conducted at the University at Buffalo. This device was a scaled prototype of the dampers of the San Bernardino County Medical Center, an isolated five-building complex in California (see Section on Principles of Friction, Viscoelastic, Yielding Steel and Fluid Viscous Dampers: Properties and Design). The test consisted of three cycles of sinusoidal displacement at frequency of 1.082 Hz and amplitude of 63 mm (peak velocity = 432 mm/s). The temperature of testing was 0, 21 and 49°C. It may be seen that the behavior of the device is nearly unaffected by temperature.

The metering tube and metering pin designs can produce force output of the type

$$F = C|\dot{u}|^2 f(u) sgn(\dot{u}) \tag{2}$$

where f(u) is a function of displacement. The design can be effective when tuned for a specific displacement signature. The pressure responsive valve design uses multiple spring loaded poppet valves to achieve a force output of the type of Equation (1). However, it has limited life due to the several moving parts in its design.

APPLICATION OF PASSIVE ENERGY DISSIPATION SYSTEMS

Passive energy dissipation systems for seismic motion reduction have been applied at the Wells Fargo Bank in San Francisco. This two-story structure suffered damage during the 1989 Loma Prieta earthquake and in 1992 it was retrofitted by the use of ADAS elements (Fierro 1993).

A two-story new school building in Phoenix, Arizona has been constructed in 1992 with is beam to column connections incorporating viscoelastic materials as shown in Figure 6. Furthermore, retrofit of the Santa Clara County Building in San Jose, California started in 1993. This 14-story steel building with exterior concrete core will have viscoelastic dampers as braces in its steel part. The Travelers Hotel, a landmark hotel built in the 1920's in Sacramento, California has been designed with fluid viscous dampers as part of its seismic retrofit scheme. Construction has not yet started.

While growth and development in this field continues, energy dissipating systems are considered for retrofit and new building and bridge applications in California and elsewhere. Currently design specifications for structures incorporating passive energy dissipating devices exist. Specifically, the Structural Engineers Association of Northern California developed "Tentative Seismic Design Requirements for Passive Energy Dissipation Systems. These requirements will eventually become part of the Uniform Building Code. The Technical Subcommittee 12 of the Building Seismic Safety Council developed provisions for "Passive Energy Dissipation Systems", which have been approved for incorporation into the 1994 NEHRP (National Earthquake Hazards Reduction Program) Recommended Provisions for the Development of Seismic Regulations for New Buildings. Furthermore, the New Technologies team of the Applied Technology Council (ATC) project No. 33 is currently developing guidelines for the design of energy dissipation systems, which will become part of the "NEHRP Recommended Guidelines for the Seismic Rehabilitation of Existing Buildings".

REFERENCES

1.   Aiken, I.D., Nims, D.K., Whittaker, A.S. and Kelly, J.M. (1993), "Testing of Passive Energy Dissipation Systems," *Earthquake Spectra, 9(3)*, 335-370.

2.   Bergman, D.M. and Goel, S.C. (1987). "Evaluation of cyclic testing of steel-plate-devices for added damping and stiffness." Report UMCE 87-10, Univ. of Michigan, Nov.

3.   British Standards Institution - BSI (1979). "Commentary on corrosion at bimetallic contacts and its alleviation." Standard PD 6484:1979, London, U.K.

4.   Chang, K.C., Soong, T.T., Lai, M.L., Neilsen, E.J. (1993), "Viscoelastic Dampers as Energy Dissipation Devices for Seismic Applications," *Earthquake Spectra*, 9(3), 371-388.

5.   Constantinou, M.C., Reinhorn, A.M., Mokha, A. and Watson, R. (1991). "Displacement control device for base-isolated bridges." Earthquake Spectra, 7(2), 179-200.

6.   Constantinou, M.C. and Symans, M.D. (1992). "Experimental and analytical investigation of seismic response of structures with supplemental fluid viscous dampers. "Report No. NCEER-92-0032, National Center for Earthquake Engineering Research, Buffalo, NY.

7.   Constantinou, M.C., Symans, M.D., Tsopelas, P. and Taylor, D.P. (1993). "Fluid viscous dampers in applications of seismic energy dissipation and seismic isolation." Proc. ATC-17-1 Seminar on Seismic Isolation, Passive Energy Dissipation, and Active Control, San Francisco, March.

8.   Fierro, E.A. and Perry, C.L. (1993). "San Francisco retrofit design using added damping and stiffness (ADAS) elements." Proc. ATC-17-1 Seminar on Seismic Isolation, Passive Energy Dissipation, and Active Control, San Francisco, March.

9.   Fitzgerald, T.F., Anagnos, T., Goodson, M. and Zsutty, T., (1989). "Slotted bolted connections in aseismic design of concentrically braced connections. " Earthquake Spectra, 5(2), 383-391.

10.   Grigorian, C.E. and Popov, E.P. (1993). "Slotted bolted connections for energy dissipation." Proc. ATC-17-1 Seminar on Seismic Isolation, Passive Energy Dissipation, and Active Control, San Francisco, March.

11.   Hsu, S-Y. and Fafitis, A., (1992). "Seismic analysis and design of frames with viscoelastic connections." J. Struct. Engrg., ASCE, 118(9), 2459-2474.

12.   Kasai, K. Munshi, J.A., Lai, M-L. and Maison,B.F. (1993). "Viscoelastic damper hysteretic model: theory, experiment, and application." Proc. ATC-17-1 Seminar on Seismic Isolation, Passive Energy Dissipation, and Active Control, San Francisco, March.

13.   Keightley, W.O. (1977). "Building damping by Coulomb friction." 6th WCEE, New Delhi, India, Jan.

14.   Keightley, W. O. (1979). "Prestressed walls for damping earthquake motions in buildings." Report of Dept. of Civil Engrg., Montana State Univ. to National Science Foundation, Sept.

15.   Kelly, J.M., Skinner, M.S. and Beucke, K.E., (1980). "Experimental testing of an energy-absorbing base isolation system. " Report No. UCB/EERC-80/35, University of California, Berkeley.

16.   Lin, R.C., Liang, Z., Soong, T.T. and Zhang, R.H. (1991), "An Experimental Study on Seismic Structural Response with Added Viscoelastic Dampers, " *Engineering Structures, 13*, 75-84.

17.   Lobo, R.F., Bracci, J.M., Shen, K.L., Reinhorn, A.M. and Soong, T.T. (1993), "Inelastic Response of R/C Structures with Viscoelastic Braces," *Earthquake Spectra*, 9 (3), 419-446.

18.   Nashif, A.D., Jones, D.I.G. and Henderson, J.P. (1985). Vibration Damping, J. Wiley and Sons, New York.

19.   Nims, D.K., Inaudi, J.A., Richter, P.J. and Kelly, J. M. (1993). "Application of the energy dissipation restraint to buildings." Proc. ATC-17-1 Seminar on Seismic Isolation, Passive Energy Dissipation, and Active Control, San Francisco, March.

20.   Tsopelas, P., Okamoto, S., Constantinou, M.C., Ozaki, D. and Fujii, S. (1994). "NCEER-TSAEI Corporation research program on sliding seismic isolation systems for bridges - experimental and analytical study of systems consisting of sliding bearings, rubber restoring force devices and fluid dampers, "Report No. 94-0002, National Center for Earthquake Engineering Research, Buffalo, NY.

21.   Whittaker, A.S., Bertero, V.V., Alonso, J.L. and Thompson, C.L. (1989). "Earthquake simulator testing of steel plate added damping and stiffness elements." Report No. UCB/EERC-89/02, University of California, Berkeley.

22.   Whitting, P.R. and Cozzarelli, F.A. (1992). "Shape memory structural dampers: material properties, design and seismic testing." Report No. NCEER-92-0013, National Center for Earthquake Engineering Research, Buffalo, NY.

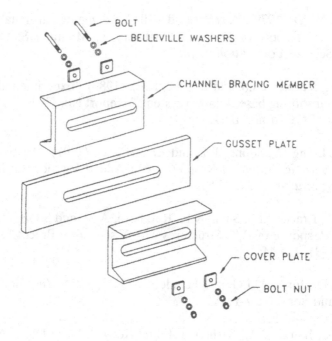

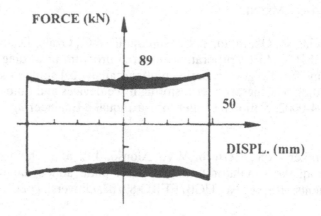

Figure 1   Slotted Bolted Connection of Fitzgerald (1989) and Typical Force-
                 Displacement Loop.

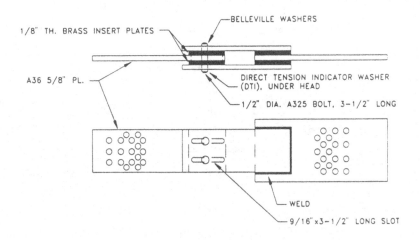

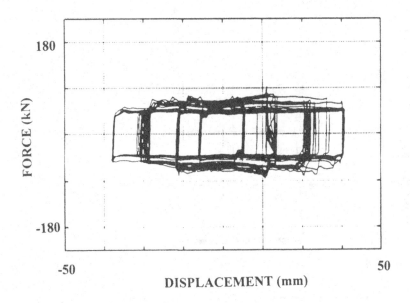

Figure 2   Slotted Bolted Connection of Grigorian (1993) and Typical Force-
                  Displacement Loop.

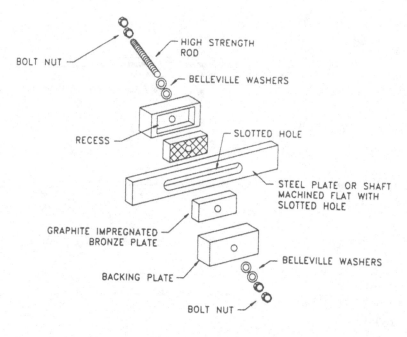

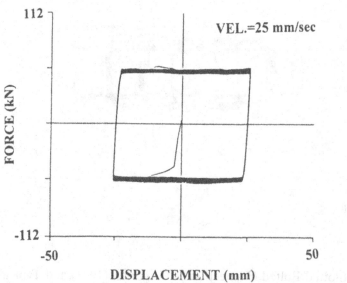

Figure 3   Friction Device of Constantinou (1991) and Force-Displacement
                Loop in 200-cycle Test.

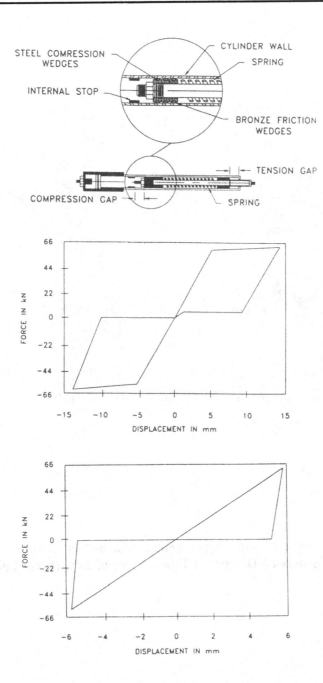

Figure 4   Energy Dissipating Restraint and Representative Force-Displacement Loops (from Nims 1993).

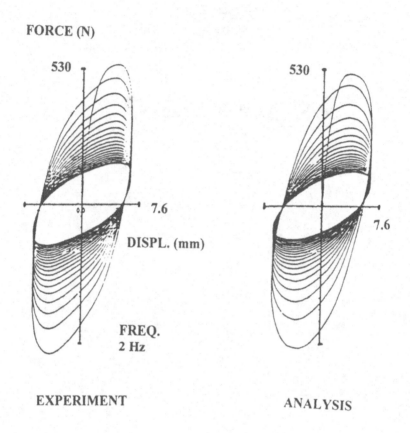

FORCE (N)

EXPERIMENT   ANALYSIS

Figure 5   Analytical and Experimental Force-Displacement Loops of
                  Viscoelastic Damper at Temperature of 24°C (from Kasai 1993).

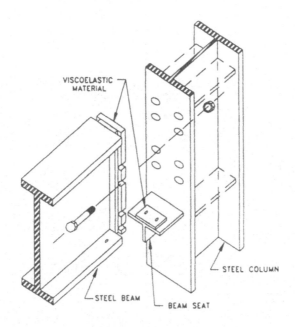

Figure 6   Detail of Beam to Column Connection with Viscoelastic Material.

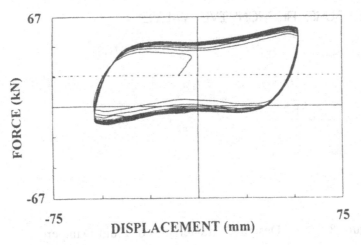

Figure 7   Force-Displacement Loop of ADAS Element (from Whittaker 1989).

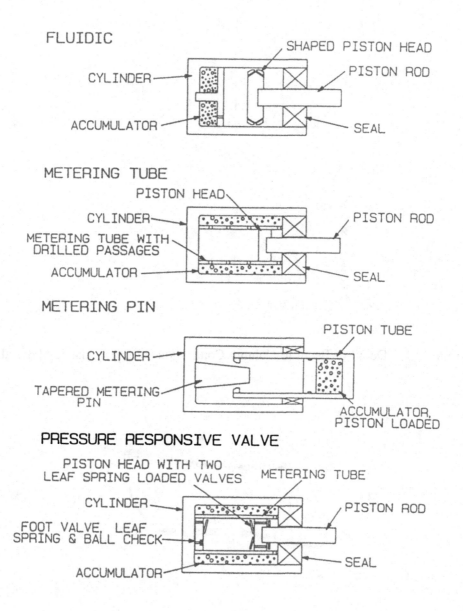

Figure 8   Design Characteristics of Fluid Dampers.

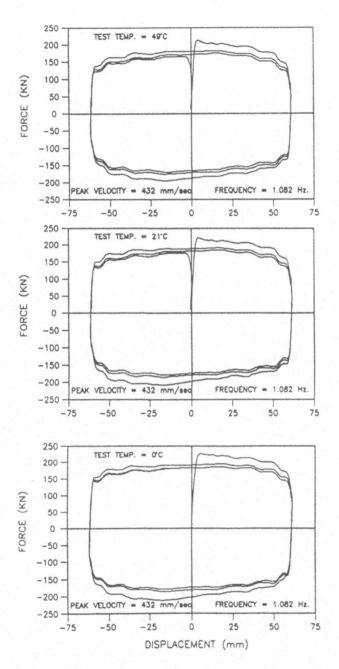

Figure 9   Force-Displacement Loops of Nonlinear Viscous Fluid Damper ($\alpha = 0.5$).

CHAPTER XIII

INTRODUCTION TO ACTIVE CONTROL

T.T. Soong

State University of New York at Buffalo, Buffalo, NY, USA

ABSTRACT

In comparison with passive energy dissipation, research and development of active structural control technology has a more recent origin. In structural engineering, active structural control is an area of research in which the motion of a structure is controlled or modified by means of the action of a control system through some external energy supply. In comparison with passive systems, a number of advantages associated with active systems can be cited; among them are (a) *enhanced effectiveness in motion control.* The degree of effectiveness is, by and large, only limited by the capacity of the control system; (b) *relative insensitivity to site conditions and ground motion;* (c) *applicability to multi-hazard mitigation situations.* An active system can be used, for example, for motion control against both strong wind and earthquakes; and (d) *selectivity of control objectives.* One may emphasize, for example, human comfort over other aspects of structural motion.

Thus motivated, considerable attention has been paid to active structural control research in recent years. It is now at the stage where actual systems have been designed, fabricated and installed in full-scale structures. A number of review articles (Miller et al., 1988; Kobori, 1988; Masri, 1988; Soong, 1988; Yang and Soong, 1988; Reinhorn and Manolis, 1989; Soong et al., 1991) and a book (Soong, 1990) have provided the reader with information and assessment on recent advances in this emerging area.

BASIC PRINCIPLES OF ACTIVE CONTROL

An active structural control system has the basic configuration as shown schematically in Fig. 1. It consists of (a) sensors located about the structure to measure either external excitations, or structural response variables, or both; (b) devices to process the measured information and to compute necessary control forces needed based on a given control algorithm; and (c) actuators, usually powered by external energy sources, to produce the required forces. When only the structural response variables are measured, the control configuration is referred to as *closed-loop control* since the structural response is continually monitored and this information is used to make continual corrections to the applied control forces. An *open-loop control* results when the control forces are regulated only by the measured excitations. In the case where the information on both the response quantities and excitation are utilized for control design, the term *open-closed loop control* is used.

To see the effect of applying such control forces to a structure under ideal conditions, consider a building structure modeled by an n-degree-of-freedom lumped mass-spring-dashpot system. The matrix equation of motion of the structural system can be written as

$$M\ddot{x}(t) + C\dot{x}(t) + Kx(t) = Du(t) + Ef(t) \tag{1}$$

where M, C and K are the $n \times n$ mass, damping and stiffness matrices, respectively, $x(t)$ is the n-dimensional displacement vector, the r-vector $f(t)$ represents the applied load or external excitation, and the m-vector u is the applied control force vector. The $n \times m$ matrix D and the $n \times r$ matrix E define the locations of the control force vector and the excitation, respectively.

Suppose that the open-closed loop configuration is used in which the control force $u(t)$ is designed to be a linear function of the measured displacement vector $x(t)$, the velocity vector $\dot{x}(t)$ and the excitation $f(t)$. The control force vector takes the form

$$u(t) = K_1 x(t) + C_1 \dot{x}(t) + E_1 f(t) \tag{2}$$

where K_1, C_1, and E_1 are respective control gains which can be time-dependent.

The substitution of equation (2) into equation (1) yields

$$M\ddot{x}(t) + (C - DC_1)\dot{x}(t) + (K - DK_1)x(t) = (E + DE_1)f(t) \tag{3}$$

Comparing equation (3) with equation (1) in the absence of control, it is seen that the effect of open-closed loop control is to modify the structural parameters (stiffness and damping) so that it can respond more favorably to the external excitation. The effect of the open-loop component is a modification (reduction or total elimination) of the excitation.

It is seen that the concept of active control is immediately appealing and exciting. On the one hand, it is capable of modifying properties of a structure in such a way as to react to external excitations in the most favorable manner. On the other hand, direct reduction of the level of excitation transmitted to the structure is also possible through active control.

The choice of the control gain matrices K_1, C_1 and E_1 in equation (2) depends on the control algorithm selected. A number of control strategies for structural applications have been developed, some of which are based on the classical optimal control theory and some are proposed for meeting specific structural performance requirements. The reader is referred to Soong (1990) for discussions of some commonly used structural control algorithms.

ORGANIZATION

The following chapters are devoted to the topic of active control. The material flows from theoretical background to practical considerations to full-scale implementation. It is designed to provide a working knowledge of this exciting and fast expanding field. Moreover, current research and world-wide development in active control technology is brought up-to-date.

REFERENCES

Kobori, T. (1988), "State-of-the-art Report: Active Seismic Response Control," *Proc. Ninth World Conference on Earthquake Engrg.*, Tokyo/Kyoto, Japan, Vol. VIII, 435-446.

Masri, S.F. (1988), "Seismic Response Control of Structural Systems: Closure," *Proc. Ninth World Conference on Earthquake Engrg.*, Tokyo/Kyoto, Japan, Vol. VIII, 497-502.

Miller, R.K., Masri, S.F., Dehghanyar, T.J. and Caughey, T.K. (1988), "Active Vibration Control of Large Civil Engineering Structures," *ASCE J. Engrg. Mech. Div.* Vol. 114, 1542-1570.

Reinhorn, A.M. and Manolis, G.D. (1989), "Recent Advances in Structural Control," *Shock and Vibration Digest*, Vol. 21, 3-8.

Soong, T.T. (1988), "State-of-the-art Review: Active Structural Control in Civil Engineering," *Engineering Structures*, Vol. 10, 74-84.

Soong, T.T. (1990), *Active Structural Control: Theory and Practice*, Longman, London, and Wiley, New York.

Soong, T.T., Reinhorn, A.M. et al. (1991), "Full Scale Implementation of Active Control - Part I: Design and Simulation," *ASCE J. Struct. Engrg.*, Vol. 117(11), 3516-3636.

Soong, T.T., Masri, S.F. and Housner, G.W. (1991) "An Overview of Active Structural Control under Seismic Loads," *Earthquake Spectra*, Vol. 7(3), 483-505.

Yang. J.N. and Soong, T.T. (1988), "Recent Advances in Active Control of Civil Engineering Structures," *J. Prob Eng. Mech.*, Vol. 3, 179-188.

CHAPTER XIV

ACTIVE CONTROL: CONCEPTS AND STRATEGIES

J. Rodellar
Technical University of Catalunya, Barcelona, Spain

ABSTRACT

This chapter describes some concepts and methods related to the design of control laws and algorithms for active control of structures. The first point will be to identify second order differential equations as models to describe the essential components of a control system. Then some issues about the *state space representation* of these models will be reviewed. State space is the mathematical framework most frequently used to formulate active control laws. *Optimal control* will be presented as a representative methodology for continuous time control. *Predictive control* will be formulated as representative of discrete time control methods, also pointing out the issue of the *time delay* and some questions about *robustness*.

This chapter does not try to be exhaustive. Control (in general) is a very wide area of knowledge and active control of structures is adopting more and more of its concepts, methods and techniques. The purpose of this chapter is to serve as an introduction to the problem of analysis and design of control laws. Many topics are left and can be covered going through the bibliography suggested in the last Section.

1. INTRODUCTION

1.1. Basic concepts and elements of an active control system

Automatic control is a branch of *system engineering* that is concerned with the design of systems able to act on a process to force it to behave according to some prescribed specifications. The behaviour of the process is described by means of a *dynamic system*, whose state is characterized by some magnitudes (*state variables*) whose time evolution depends on the interaction with the environment and on the internal interactions. The relationship between the system and the external world can be understood in terms of *input* and *output* variables. The outputs are measurable magnitudes representing the system's response. The inputs represent external actions on the system and can be classified in *control* and *perturbation* variables. While perturbations are not manipulable and usually correspond to natural excitations, controls are manipulable actions. We call *controller* to that system designed to manipulate the inputs to make the system's output to reach a prescribed value (*set point*) according to some desired performance.

We can distinguish two controller arrangements as illustrated in Figure 1: *open loop* and *closed loop*. In the first one, the controller manipulates the input with the purpose of driving the output to the setpoint, but no use is made of the resulting output. This control action can be generated according to a prescribed schedule based on previous experience or even a methodology. But there is no possibility to compensate for errors due to perturbations or any other effect in real time unless accurate information on them is known. Thus open loop is not a general and common procedure for control. Closed loop control uses the system response as *feedback* information, with the consequent possibility of compensating effects influencing on the response in a negative way. The control action can be *manual*, when some human operator is involved in its implementation, or *automatic* in the case that only artificial devices are involved in the control loop.

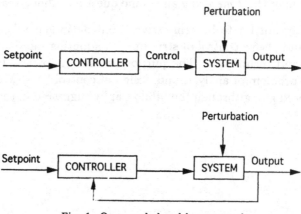

Fig. 1.–Open and closed loop control.

Active structural control systems lie in the framework described above. Essentially, they emerge when structures are viewed as dynamic systems whose behaviour can be modified by an automatic control system using similar principles to those used in the control of other more "controllable" systems, as those in industrial, aerospace, mechanical and many other areas.

In order to translate this idea into a practical system, we need to make clear how to identify and realize input and output variables and to build the controller. Concerning the outputs, we need *sensors* to measure response characteristics of the structure, such as displacements, velocities, accelerations, etc. The perturbation inputs are due to external excitations, produced by earthquakes, wind, impacts, or any other vibration source. The control inputs have to be applied by appropriate *actuators* in the form of forces, torques, accelerations, etc. In other parts of this book different kind of actuators have been described. Here we will concentrate on the design of the controller.

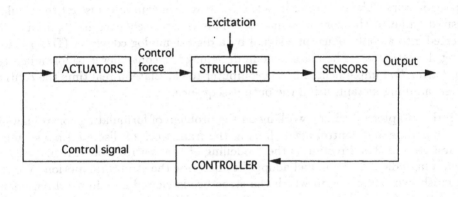

Fig. 2.–Basic elements of an active control loop.

The *controller* can be considered as an information processor closing the loop as indicated in Figure 2. Its role is to supply the signal (sent to drive the actuators) as a feedback function of the signal coming from the sensors. When we are working in the design, the controller can be considered as a mathematical object to be formulated within the framework of a *control theory*. From implementation point of view, the controller can be *analogous* or *digital*. In the first case, all the devices work in a continuous way and, consequently, the controller can be realized by means of components which can directly implement a *feedback control law* formulated in *continuous time*. In the second case, the controller is based on a digital computer as the main component. This is the most common situation nowadays in the control applications, due to the spectacular advances of the microelectronics and the advantages of digital technology in front of the analogous one. Figure 3 shows the control loop in this case.

Now the continuous time signal from the sensors is sampled by an analog–

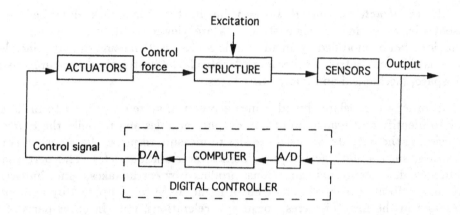

Fig. 3.–Basic elements of an active digital control loop.

digital (A/D) converter, thus supplying a sequence of measured output values at each sampling instant. The computer uses the value at each sampling instant to calculate a desired value of the control sequence. This control sequence is required to be converted into a continuous time signal by a digital–analog converter (D/A) before feeding the actuators. The main issue in designing the controller is to formulate the *control algorithm* implemented in the computer to calculate in real time the control, at each sampling instant, using the output sequence.

In this chapter we will be working on the problem of formulating control laws in continuous time and control algorithms in the framework of discrete time systems. The first step in this direction is the modelling of the control loop. In this respect, the main ingredient is the model adopted to describe the structural motion. We will distinguish two cases: one in which the structure is viewed as a lumped parameter system; the other, in which the structure is modelled as a distributed parameter system. In both cases we will try to derive a common model to be the basis of further steps in the formulation of the control problem.

1.2. Modelling of the control loop for a lumped parameter structure

Trying to be didactic in the presentation, we will consider a particular example. Figure 4 shows a block diagram with the basic elements of an experimental control loop. The structure is a 1:4 scale, three bays, six stories metal model set up in the Department of Civil Engineering, University of New York at Buffalo (USA). The structure and the additional masses placed at the different floors weights 19.18 tons. Output feedback signals to the control system are the displacements and velocities of each floor relative to the ground. These signals are sampled and converted to discrete-time values by A/D converters. These values are used by a computer to calculate the desired values of the control sequences which are translated into analog control signals by D/A converters. These control signals feed the servovalves of two tendon controllers, placed between the ground and first floor and between the second and third floors respectively, which apply horizontal control forces on the structure at floors 1, 2 and 3.

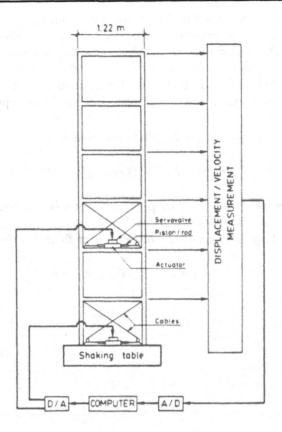

Fig. 4.–Experimental control loop.

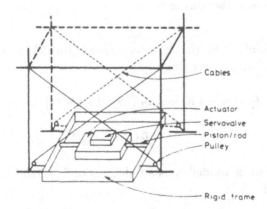

Fig. 5.–Active tendon controller.

Figure 5 shows a diagram of a tendon controller. It is composed by four cables and an hydraulic actuator. The cables are braced to the upper floor by one of their ends while the other ends are attached to a horizontal rigid frame through four pulleys. The frame is connected to the piston rod of an hydraulic actuator whose motion is commanded by a servovalve proportionally to the difference between the analog signal from the D/A converter and the actual displacement of the piston/rod. In this way the tensions of the cables are actively modified, which results in horizontal control forces on the structure. In fact, consider the actuator 1 at instant t when the first floor relative displacement is $d_1(t)$. Tensions at cables, as represented in Figure 6, verify

$$T_{1d} = T_0 + K_t \left[d_1(t) \cos \alpha_1 + u_1(t) \right] \tag{1a}$$

$$T_{1q} = T_0 - K_t \left[d_1(t) \cos \alpha_1 + u_1(t) \right] \tag{1b}$$

where T_0 is a pretension in order to prevent the tension release during control application. K_t is the stiffness of the cables and u_1 is the displacement of the actuator. The resulting horizontal control force f_1 at floor 1 is then

$$f_1(t) = 4 K_t \cos \alpha_1 \left[d_1(t) \cos \alpha_1 + u_1(t) \right] \tag{2}$$

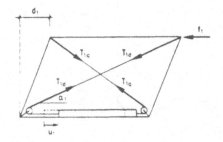 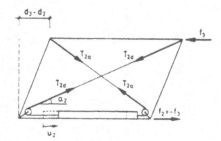

Fig. 6.–Control force applied on the structure by the actuator 1.   Fig. 7.–Control forces applied on the structure by the actuator 2.

The same analysis for the actuator 2 (see Figure 7) shows the presence of control forces at floors 2 and 3 verifying:

$$f_2(t) = -4 K_t \cos \alpha_2 \left[\{ d_3(t) - d_2(t) \} \cos \alpha_2 + u_2(t) \right] \tag{3a}$$

$$f_3(t) = -f_2(t) \tag{3b}$$

The structure can be modelled as a 6 degrees of freedom lumped mass model whose motion is described by

$$M \ddot{d} + C \dot{d} + K d = -f(t) - M j \ddot{d}_0(t) \tag{4}$$

where M, C and K are the mass, damping and stiffness matrices, d the floor displacement vector (relative to the ground), j the unit vector and f the control force vector $f^T = [f_1, f_2, f_3, 0, 0, 0]$ which can be written, according to eqs. (2) and (3), in the form:

$$f = K_p d + L u \qquad (5)$$

where

$$K_p = \begin{bmatrix} 4\,K_t \cos^2 \alpha_1 & 0 & 0 & 0 & 0 & 0 \\ 0 & 4\,K_t \cos^2 \alpha_2 & -4\,K_t \cos^2 \alpha_2 & 0 & 0 & 0 \\ 0 & -4\,K_t \cos^2 \alpha_2 & 4\,K_t \cos^2 \alpha_2 & 0 & 0 & 0 \\ 0 & 0 & 0 & 0 & 0 & 0 \\ 0 & 0 & 0 & 0 & 0 & 0 \\ 0 & 0 & 0 & 0 & 0 & 0 \end{bmatrix} \qquad (6a)$$

$$L = \begin{bmatrix} 4\,K_t \cos^2 \alpha_1 & 0 \\ 0 & -4\,K_t \cos^2 \alpha_2 \\ 0 & 4\,K_t \cos^2 \alpha_2 \\ 0 & 0 \\ 0 & 0 \\ 0 & 0 \end{bmatrix} \qquad (6b)$$

By substituting (5) into (4) one can write

$$M\,\ddot{d} + C\,\dot{d} + K'\,d = -L\,u(t) - M\,j\,\ddot{d}_0(t) \qquad (7)$$

where

$$K' = K + K_p \qquad (8)$$

Equation (7) describes the motion of the structure under control by active cables. One can see the effect of the cables is double: they supply a passive control action $K_p\,d$ and an active control $L\,u$. The first one is equivalent to a change in the stiffness matrix of the structure which is now K'. The active control action is represented by $L\,u$ and is produced by the displacement u of the actuators as commanded in closed loop by the control system. The controller design now consists of formulating the control law giving u as a feedback of the structural response.

1.3. Modelling of the control loop for a distributed parameter structure

Again we will work with an example. Consider a simple–span bridge model with constant flexural rigidity EI, mass per unit length m, damping c and span L. Assume there is a distributed force $F(x, t)$ exciting the structure, where x is the distance measured from the left support. Assume we can design a distributed control force $U(x, t)$. The equation describing the vertical motion y of the structure is

$$m\frac{\partial^2 y}{\partial t^2} + c\frac{\partial y}{\partial t} + EI\frac{\partial^4 y}{\partial x^4} = F(x, t) + U(x, t) \qquad (9)$$

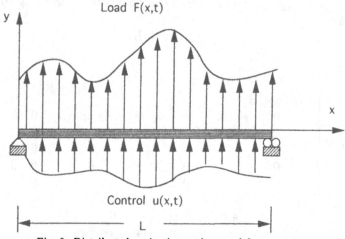

Fig. 8.–Distributed excitation and control forces.

We can assume the solution of this equation in the form

$$y(x,t) = \sum_{j=1}^{\infty} \Phi_j(x)\eta_j(t) \qquad (10)$$

where $\eta_j(t)$ is the generalized coordinate of mode j and $\Phi_j(x)$ is the characteristic function of this mode which satisfies the boundary conditions. In this example, it is

$$\Phi_j(x) = \sqrt{\frac{2}{L}} sin\frac{i\pi x}{L}$$

Substitution of (10) into (9) and applying a known integral transform leads to the following set of ordinary second order differential equations

$$\ddot{\eta}_i + 2\nu_i\omega_i\dot{\eta}_i + \omega_i^2\eta_i = f_i(t) + u_i(t)$$

$$i = 1,\cdots,\infty \qquad (11)$$

where

$$\omega_i = \frac{i^2\pi^2}{L^2}\sqrt{\frac{EI}{m}}; \qquad \nu_i = \frac{c}{2m\omega_i}$$

and

$$f_i(t) = m^{-1}\int_0^L F(x,t)\Phi_i(x)dx$$

$$u_i(t) = m^{-1}\int_0^L U(x,t)\Phi_i(x)dx \qquad (12)$$

which can be called *modal excitation* and *modal control* forces respectively. Using this kind of model, we are in the framework of the so–called *modal control*, in which the control problem can be stated in the following terms: design a control law to generate the modal forces u_i using, as feedback signals, the modal coordinates η_j. Some problems arise from the fact that:

o In principle, there exists an infinite number of modes.

o Modal coordinates do not correspond to physical variables that can be measured by sensors.

o Modal control forces are not real forces directly implementable by actuators.

The first point motivates the consideration of a reduced finite number of modes. The second one suggests the need of some algorithm to estimate modal coordinates from the response measured at specific points. The third point requires the design of specific actuators.

Besides this problems, we can see that (11) is a set of second order ordinary differential equations similar to the one derived in (7) for a lumped mass structure. This structural equation is the base for the controller design.

2. STATE SPACE REPRESENTATION

2.1 State equation for the structural motion

As illustrated before, in many structural control problems, the structure is described by a spatially discretized linear model with n degrees of freedom described by the following equation of motion:

$$M\ddot{d}(t) + C\dot{d}(t) + Kd(t) = L_e f(t) + L_c u(t) \tag{13}$$

where M, C and K are, respectively, the mass, damping and stiffness matrices, d is the displacement vector, f is the $m \times 1$ dynamic excitation vector and u is the $r \times 1$ control vector. L_e, L_c are matrices indicating the location of the control actions and excitations.

Equation (13) can be written in a first order form

$$\dot{x}(t) = Fx(t) + G_c u(t) + G_e f(t) \tag{14}$$

where x is the $2n \times 1$ *state vector*, F is the $2n \times 2n$ *system matrix*

$$x(t) = \begin{pmatrix} d(t) \\ \dot{d}(t) \end{pmatrix}; \qquad F = \begin{pmatrix} 0 & I \\ -M^{-1}K & -M^{-1}C \end{pmatrix} \tag{15}$$

and G_c, G_e are $2n \times r$ and $2n \times m$ location matrices

$$G_c = \begin{pmatrix} \mathbf{0} \\ M^{-1}L_c \end{pmatrix}; \qquad G_e = \begin{pmatrix} \mathbf{0} \\ M^{-1}L_e \end{pmatrix} \qquad (16)$$

(14) is the so-called linear time–invariant state equation. It is a particular case of a more general description of dynamic systems represented by a first order matrix differential equation of the form

$$\dot{x} = g[x(t), u(t), t] \qquad (17)$$

Many of the modern control laws and algorithms are formulated based on equation (14). Therefore, we will concentrate here on presenting some of the most relevant issues about this equation, which are frequently used in control problems.

2.2. Solution of the state equation

Let us call

$$v = G_c u + G_e f$$

With the initial condition $x(t_0) = x_0$, the analytical solution of (2) is

$$x(t) = \exp[(t - t_0)F]x_0 + \int_{t_0}^{t} \exp[(t - \tau)F]v(\tau)\, d\tau \qquad (18)$$

where, for a square matrix θ, exp denotes the *exponential matrix* defined as the convergent series

$$\exp(\theta) = \sum_{i=0}^{\infty} \frac{\theta^i}{i!} = I + \frac{\theta}{1!} + \frac{\theta^2}{2!} + \cdots$$

The first term in (18) depends only on the initial state, so representing the free response. The second one is a convolution term and represents the forced response due to the input.

2.3. Discretization of the state equation

Consider now the time discretized with a sampling period T. We may obtain a discrete time model from the continuous time equation (14) by applying the solution (18) between two consecutive time instants. To do this, we can substitute $t_0 = kT$ and $t = kT + T$ into (18) and write

$$x(kT + T) = \exp(TF)x(kT) + \int_{kT}^{(k+1)T} \exp\{[(k+1)T - \tau]F\}v(\tau)\, d\tau \qquad (19)$$

To solve the integral involved in (19), the continuous–time evolution of $v(\tau)$ in the discretization interval is required. If it is considered that the excitation vector is known only at discrete instants, $v(\tau)$ can be defined by interpolating the discrete values. A common simple criterion may consist of assuming a zero order interpolation, by which the sampled value at instant k is kept constant over the discretization interval; i.e.

$$v(\tau) = v(kT) \qquad kT \leq \tau < (k+1)T \qquad (20)$$

By substituting (20) into (19), the following discrete - time equation is obtained:

$$x(kT+T) = Ax(kT) + Bu(kT) + Pf(kT) \qquad (21)$$

where A, B and P are matrices given by

$$A = \exp(TF); \qquad B = F^{-1}(A-I)G_c; \qquad P = F^{-1}(A-I)G_e \qquad (22)$$

The system matrix F is always non-singular, its inverse being

$$F^{-1} = \begin{pmatrix} -K^{-1}C & -K^{-1}M \\ I & 0 \end{pmatrix}$$

The recursive application of (21), starting from the initial conditions, allows the computation of the state vector x at each time instant. Also eq. (21) gives a description of the closed loop in a digital implementation, since the signals converters (A/D–D/A) perform the operations of sampling and zero–order interpolation in a synchronized manner. Therefore, the discrete–time state model formulated above can be used in two ways: (i) for the simulation of the structural response under excitations and control actions; (ii) as base model for the formulation of discrete–time control laws (algorithms).

The numerical evaluation of the exponential matrix $A = \exp(TF)$ is the key point of this discretization procedure. There are different methods to perform this computation and a number of software packages are available.

2.4. Stability

Stability is probably the most important property that a control system has to preserve, since it is concerned with the capability of the system to exhibit a solution bounded and close to the setpoint. There are different ways to study the stability of a system. Here we will consider two ways: one is concerned with the maintenance of an equilibrium position with no input (*equilibrium stability*); the other is concerned with the boundedness of the response for bounded inputs (*external stability*). Now we give the definitions for both continuous time and discrete time systems.

Equilibrium stability. Definitions

Be a system without inputs described by

$$\dot{x}(t) = g[x, 0, t] \qquad x(t_0) = x_0 \tag{23}$$

or (in discrete time)
$$x(k+1) = g[x(k), 0, k] \qquad x(k_0) = x_0 \tag{24}$$

x_e is an *equilibrium state* of the system if the system's state at t_0 is x_e and $x(t) = x_e$ for all $t \geq t_0$.

Be the origin an equilibrium state

* The origin (for system (23)) is *stable* (in the sense of Lyapunov) if, for any t_0 and any $\varepsilon > 0$, there exists a real number $\delta(\varepsilon, t_0)$ such that $\|x(t_0)\| < \delta$ implies that $\|x(t)\| < \varepsilon$ for all $t \geq t_0$.

* The origin (for system (23)) is *asymptotically stable* if it is stable and, moreover, if for any t_0, there exists $\delta > 0$ (possibly dependent on t_0) such that $\|x(t_0)\| < \delta$ implies that $\lim_{t \to \infty} \|x(t)\| = 0$. If, moreover, $\delta(t_0)$ can be arbitrarily large, the origin is called *globally* asymptotically stable.

When, in any of the two cases above, the scalar δ does not depend on the initial time t_0, the word *uniform* is added to the stability definitions.

For the discrete–time system (24) equivalent definitions hold considering instant k instead of time t.

External stability. Definition

Be a forced system of the form

$$\dot{x}(t) = f[x, u, t] \qquad x(t_0) = x_0 \tag{25}$$

or

$$x(k+1) = f[x(k), u(k), k] \qquad x(k_0) = x_0 \tag{26}$$

* The system (25) is *externally stable* if for any t_0, $x(t_0)$ and $u(t)$ such that $\|u(t)\| \leq \delta$ for all $t \geq t_0$, there exists a scalar ε (possibly dependent on t_0, $x(t_0)$ and δ) such that $\|x(t)\| \leq \varepsilon$ for all $t \geq t_0$. Analogously for the discrete time system (26).

The above condition can be stated in the following equivalent words: if any bounded input produces a bounded state vector.

Now we specialize in time invariant linear systems to derive practical criteria to check the stability.

Stability criteria for continuous time systems

Be the unforced system described by the state equation

$$\dot{x}(t) = F\,x(t)$$
$$x(t_0) = x_0 \tag{27}$$

with solution

$$x(t) = \exp[F(t - t_0)]x_0 \tag{28}$$

Be $\lambda_i = \alpha_i + j\mu_i \quad (i = 1, \cdots, n)$ the eigenvalues of F.

There exists a set of n independent vectors defining a transformation matrix T that allows to write eq. (27) in the form

$$\dot{q}(t) = J q(t)$$
$$q(t_0) = q_0 \tag{29}$$

with $x = Tq$ and $J = T^{-1}FT$. The solution of (29) has the form

$$q(t) = \exp[J(t - t_0)]q_0 \tag{30}$$

There is one case in which J becomes diagonal:

$$J = \begin{pmatrix} \lambda_1 & & & & \\ & \ddots & & & \\ & & \lambda_i & & \\ & & & \ddots & \\ & & & & \lambda_n \end{pmatrix} \tag{31}$$

In such a case, eq.(29) is equivalent to a set of n independent scalar equations with solution

$$q_i(t) = \exp[\lambda_i(t - t_0)]q_i(t_0) \tag{32}$$

Now is clear how stability depends on the real part α_i of the eigenvalue λ_i. In fact

$$\alpha_i > 0 \Rightarrow \lim_{t \to \infty} q_i(t) = \infty \Rightarrow instability$$

$$\alpha_i = 0 \Rightarrow q_i(t) = \exp[j\mu_i(t - t_0)]q_i(t_0) \neq 0\ bounded \Rightarrow$$
$$\Rightarrow Lyapunov\ \ stability$$

$$\alpha_i < 0 \Rightarrow \lim_{t \to \infty} q_i(t) = 0 \Rightarrow asymptotic\ stability$$

If matrix J is not diagonal, it can be decomposed into a number $s < n$ of Jordan blocks J_i in the form

$$J = \begin{pmatrix} J_1 & & & & \\ & \ddots & & & \\ & & J_i & & \\ & & & \ddots & \\ & & & & J_s \end{pmatrix} \tag{33}$$

which verifies

$$\exp[J(t{-}t_0)] = \begin{pmatrix} \exp[J_1(t{-}t_0)] & & & & \\ & \ddots & & & \\ & & \exp[J_i(t{-}t_0)] & & \\ & & & \ddots & \\ & & & & \exp[J_s(t{-}t_0)] \end{pmatrix} \tag{34}$$

In this case, eq. (29) can be split into s independent vector equations with solution

$$q_i(t) = \exp[J_i(t - t_0)]\, q_i(t_0) \tag{35}$$

The exponential matrices $\exp[J_i(t - t_0)]$ have a closed form

$$\exp[J_i(t - t_0)] = \exp[\lambda_i(t - t_0)]\, \Theta_i(t) \tag{36}$$

where $\Theta_i(t)$ is a matrix with entries increasing with time in the form of powers.

As an example, we can consider the case of a Jordan block of the form

$$J_i = \begin{pmatrix} \lambda_i & 0 & 0 & 0 \\ 1 & \lambda_i & 0 & 0 \\ 0 & 1 & \lambda_i & 0 \\ 0 & 0 & 1 & \lambda_i \end{pmatrix} \tag{37}$$

It is an exercise to verify that

$$\exp[J_i(t - t_0)] = \exp[\lambda_i(t - t_0)] \begin{pmatrix} 1 & 0 & 0 & 0 \\ t & 1 & 0 & 0 \\ \frac{t^2}{2!} & t & 1 & 0 \\ \frac{t^3}{3!} & \frac{t^2}{2!} & t & 1 \end{pmatrix} \tag{38}$$

Concerning the stability, it is readily seen now in eq. (35) how it depends on α_i. In fact,

$$\alpha_i \geq 0 \Rightarrow \textit{instability}$$
$$\alpha_i < 0 \Rightarrow \textit{asymptotic stability}$$

In summary, the equilibrium stability conditions for the linear system (27) can be stated in Table 1.

For the case of *external stability*, it can be shown that if the system (27) is asymptotically stable, it is also externally stable when forced with inputs entering in the form $Gu(t)$.

TABLE 1: Stability criteria for
continuous time invariant linear systems

Instability	If $\alpha_i > 0$ for any eigenvalue with Jordan block of size 1 If $\alpha_i \geq 0$ for any eigenvalue with Jordan block of size >1
Stability in the sense of Lyapunov	If $\alpha_i \leq 0$ for any eigenvalue with Jordan block of size 1 If $\alpha_i < 0$ for any eigenvalue with Jordan block of size >1
Asymptotic stability	If $\alpha_i < 0$ for any eigenvalue

Stability criteria for discrete time systems

Be the unforced discrete time system

$$x(k+1) = Ax(k)$$
$$x(0) = x_0$$

(39)

It has a closed solution of the form

$$x(k) = A^k x_0$$

(40)

Be $\lambda_i = |\lambda_i| \exp(j\omega_i)$ $(i = 1, \cdots, n)$ the eigenvalues of matrix A and T the associated transformation matrix that allows to write (39) in the form

$$q(k+1) = Jq(k)$$

(41)

$$q(0) = q_0$$

where $q = Tx$ and $J = T^{-1}FT$. The solution of (41) is

$$q(k) = J^k q_0 \tag{42}$$

If matrix J is diagonal like (31), eq. (39) is decomposed into n independent scalar equations with solution

$$q_i(k) = \lambda_i^k q_{i0} = |\lambda_i| \exp(j\omega_i k) q_{i0} \tag{43}$$

whose stability depends on the modulus of the eigenvalue λ_i. In fact,

$$|\lambda_i| > 1 \Rightarrow \lim_{k \to \infty} q_i(k) = \infty \Rightarrow instability$$

$$|\lambda_i| = 1 \Rightarrow q_i(k) = \exp(j\omega_i k) q_{i0} \neq 0 \; bounded \Rightarrow$$

$$\Rightarrow Lyapunov \quad stability$$

$$|\lambda_i| < 1 \Rightarrow \lim_{k \to \infty} q_i(k) = 0 \Rightarrow asymptotic \quad stability$$

If J is not diagonal, it decomposes into a number $s < n$ of Jordan blocks J_i like in (33), verifying

$$J^k = \begin{pmatrix} J_1^k & & & & \\ & \ddots & & & \\ & & J_i^k & & \\ & & & \ddots & \\ & & & & J_s^k \end{pmatrix} \tag{44}$$

In this case, eq. (39) decomposes into s independent vectorial equations with solution

$$q_i(k) = J_i^k q_{i0} \tag{45}$$

Matrices J_i^k have a closed expression

$$J_i^k = (\lambda_i I + \Theta_i)^k = \sum_{r=0}^{k} \frac{k!}{r!(k-r)!} \lambda_i^{k-r} \Theta_i^r \tag{46}$$

where λ_i is the eigenvalue associated to the block J_i and Θ_i is a constant matrix with entries either 0 or 1. By the way of example, if we consider a block of size $n_i \times n_i$

$$J_i = \begin{pmatrix} \lambda_i & 0 & 0 & \cdots & 0 & 0 \\ 1 & \lambda_i & 0 & \cdots & 0 & 0 \\ 0 & 1 & \lambda_i & \cdots & 0 & 0 \\ \vdots & \vdots & \vdots & \cdots & \vdots & \vdots \\ 0 & 0 & 0 & \cdots & 1 & \lambda_i \end{pmatrix} \tag{47}$$

we can readily verify that expression (46) holds with

$$
\boldsymbol{\Theta}_i = \begin{pmatrix}
0 & 0 & 0 & \cdots & 0 & 0 \\
1 & 0 & 0 & \cdots & 0 & 0 \\
0 & 1 & 0 & \cdots & 0 & 0 \\
\vdots & \vdots & \vdots & \cdots & \vdots & \vdots \\
0 & 0 & 0 & \cdots & 1 & 0
\end{pmatrix}
$$

Using (46), it is easy to observe the influence of the eigenvalue λ_i in the stability of the solution (45). In fact,

$$|\lambda_i| \geq 1 \Rightarrow instability$$

$$|\lambda_i| < 1 \Rightarrow asymptotic\ \ stability$$

Table 2 summarizes the stability condition for system (39).

It can be shown that, if the system (39) is asymptotically stable, it is also *externally stable* when forced with inputs entering in the form $\boldsymbol{B}\boldsymbol{u}(k)$.

TABLE 2:   Stability criteria for
discrete time invariant linear systems

Instability	If $|\lambda_i| > 1$ for any eigenvalue with Jordan block of size 1 If $|\lambda_i| \geq 1$ for any eigenvalue with Jordan block of size > 1
Stability in the sense of Lyapunov	If $|\lambda_i| \leq 1$ for any eigenvalue with Jordan block of size 1 If $|\lambda_i| < 1$ for any eigenvalue with Jordan block of size > 1
Asymptotic stability	If $|\lambda_i| < 1$ for any eigenvalue

3. OPTIMAL CONTROL

3.1. Generic problem and solution framework

Be a dynamic system described by a state equation of the form

$$\dot{x} = f[x(t), u(t), t]$$
$$x(t_0) = x_0 \tag{48}$$

where x is the $n \times 1$ *state vector* and u is the $r \times 1$ *control vector*. Both $x(t), u(t)$ are continuous functions and f is assumed to have continuous partial derivatives with respect to both x, u. Then, given the initial condition x_0 and the control vector $u(t)$ over an interval $[t_0, t_f]$, there exist a unique continuous trajectory $x(t)$ solution of (48).

Define the scalar performance index

$$J = \theta[x(t_f), t_f] + \int_{t_0}^{t_f} \Phi[x(t), u(t), t] dt \tag{49}$$

where θ, Φ have continuous partial derivatives with respect to both x, u, and t_0, t_f are fixed instants.

We consider the following optimal control problem: find the control $u(t)$ such that the performance index (49) is minimized with the restriction given by eq. (48).

Other constraints can be also considered in a general case, for example, limits on the state and the control. However we will not consider this case for the sake of simplicity.

We will follow the main steps to solve the above problem in the framework of variational calculus, or the so-called *hamiltonian formulation*. First, eqs. (48)–(49) are adjoined introducing a $n \times 1$ Lagrange multiplier vector function $\lambda(t)$ in the form

$$J = \theta[x(t_f), t_f] + \int_{t_0}^{t_f} \{\Phi[x(t), u(t), t] + \lambda^T(t)[f(x(t), u(t), t) - \dot{x}(t)]\} dt \tag{50}$$

The Hamiltonian H is introduced as

$$H[x(t), u(t), \lambda(t), t] = \Phi[x(t), u(t), t] + \lambda^T(t) f[x(t), u(t), t] \tag{51}$$

Introducing H in (50) and integrating by parts the term

$$\int_{t_0}^{t_f} \lambda^T \dot{x} dt,$$

we may write

$$J = \theta[x(t_f), t_f] - \lambda(t_f)x(t_f) + \lambda(t_0)x(t_0) + \int_{t_0}^{t_f} [H + \dot{\lambda}^T x] dt \qquad (52)$$

Now we can consider arbitrary variations $\delta x(t), \delta u(t)$ with respect to the optimal state and control trajectories, respectively. These variations produce a variation on the optimal value of the performance index J. A necessary condition for a minimum is that the first variation in J is zero. Taking the first variation of (52) (using the fact that $x(t_0) = x_0$ is fixed), it results

$$\delta J = [\frac{\partial \theta}{\partial x} - \lambda]^T \delta x|_{t=t_f} + \int_{t_0}^{t_f} \{[\frac{\partial H}{\partial x} + \dot{\lambda}^T]\delta x + \frac{\partial H}{\partial u} \delta u\} dt \qquad (53)$$

Imposing $\delta J = 0$, we have

$$\lambda(t_f) = \frac{\partial \theta}{\partial x}(t_f) \qquad (54)$$

$$\dot{\lambda} = -\frac{\partial H}{\partial x} \qquad (55)$$

$$\frac{\partial H}{\partial u} = 0 \qquad (56)$$

Eqs. (54)–(56), together with the system equation and the initial condition in (48), define a two–point boundary value problem. This problem gives a framework including a wide family of problems. For each case, these equations have to be solved to obtain the control that can produce the minimum value of index J. Sufficient conditions involve second variations and are difficult to state in the general case, but they can be drawn for specific examples, depending on the system equation and the performance index considered in every case.

In active control of structures, system (48) is usually linear and time invariant and the performance index is quadratic in the state and the control. In the next section we particularize the previous framework for this case.

3.2. Linear quadratic optimal control

Consider the system

$$\dot{x} = Fx + Gu; \qquad x(t_0) = x_0 \qquad (57)$$

and the performance index

$$J = \frac{1}{2}x(t_f)^T Sx(t_f) + \frac{1}{2}\int_{t_0}^{t_f} \{x(t)^T Q(t)x(t) + u(t)^T R(t)u(t)\} dt \qquad (58)$$

S, Q, R are *weighting* matrices and their values are selected depending on the relative importance given to the different terms in their contribution to the performance index J. S penalizes the state at the final time t_f. Big values of Q represent the desire of keeping the state vector close to the origin during the minimization interval $[t_0, t_f]$, while big values of R require the control not to be excessive. Therefore, playing with the relative values of Q and R, it is possible to design controllers with a compromise between both requirements.

The Hamiltonian now is

$$H[x(t), u(t), \lambda(t), t] = \frac{1}{2} x^T Q x + \frac{1}{2} u^T R u + \lambda^T F x + \lambda^T G u \tag{59}$$

Eqs. (54)–(56) take the form $\lambda(t_f) = \dfrac{\partial \theta}{\partial x}(t_f) = S x(t_f)$ $\tag{60}$

$$\frac{\partial H}{\partial x} = -\dot{\lambda} = Q(t) x(t) + F^T \lambda(t) \tag{61}$$

$$\frac{\partial H}{\partial u} = 0 = R(t) u(t) + G^T \lambda(t) \tag{62}$$

Assume the following relation

$$\lambda(t) = P(t) x(t) \tag{63}$$

where P is an unknown matrix.

Substitution of (63) into (62) yields

$$u(t) = -R^{-1}(t) G^T P(t) x(t) \tag{64}$$

Substitution of (63) and (64) into (57) results in

$$P(t) = -P(t) F + P(t) G R^{-1}(t) G^T P(t) - F^T P(t) - Q(t) \tag{65}$$

while the terminal condition (54) is

$$\lambda(t_f) = S \tag{66}$$

Eq. (65) is the so called matrix *Riccati equation*, very well known in control theory. Using the condition (66), it can be solved backwards in time from the terminal time t_f until the initial time t_0. Then, if the time variant gain matrix $D(t) = R^{-1}(t) G^T P(t)$ is stored at each time t, the closed loop control law is

$$u(t) = -D(t) x(t) \tag{67}$$

It is known in control theory that weighting matrices S, Q have to be at least positive semidefinite and R positive definite for J to reach a minimum with the control law (67). Also they are usually selected symmetric.

From an implementation point of view, the requirement of solving the matrix Riccati equation and storing the time varying solution for its further use in real time can be unpractical. Fortunately, it has been proved that, in the case that $S = 0$ and Q, R are time invariant, there exists $\lim_{t \to \infty} P(t) = \bar{P}$, $\bar{P}$ being the solution of the algebraic matrix Riccati equation

$$0 = -\bar{P}F + \bar{P}GR^{-1}G^T\bar{P} - F\bar{P} - Q \tag{68}$$

This can be interpreted considering that, when we solve eq. (55) backwards from t_f to t_0, the elements of matrix $P(t)$ have a profile like the one illustrated in Figure 9. There is a "transient" stage followed by a "steady" stage in which matrix $P(t)$ tends to $\bar{P}$. If the interval $[t_0, t_f]$ is long enough, we can approximate $P(t)$ by $\bar{P}$ for all time. In this case, the control law is

$$u(t) = -Dx(t) \tag{69}$$

where now the gain matrix is constant

$$D = R^{-1}G^T\bar{P} \tag{70}$$

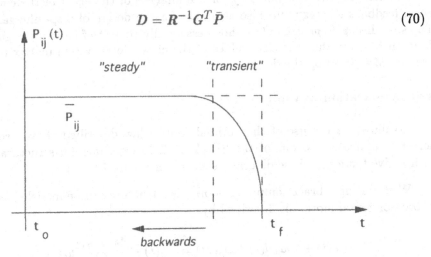

Fig. 9.–Typical element of the Riccati matrix.

This control law is the one used in most of the practical applications of optimal control. Now the gain matrix is calculated once prior to the implementation after selecting the weighting matrices, which are the parameters to be selected by the designer.

3.3. Gain matrix and eigenvalues

As we have seen previoulsy, eigenvalues of the system matrix F determine the stability properties of a linear system like the one described by eq. (57). Moreover these eigenvalues determine the system dynamics, in particular the modal frequency and damping characteristics. If model (57) is a description of a structural system like the one in (13), the eigenvalues are arranged in complex conjugate pairs of the form

$$\lambda_i = \zeta_i \omega_i \pm j\omega_i \sqrt{1 - \zeta_i^2}, \qquad j = \sqrt{-1} \tag{71}$$

ω_i, ζ_i being the modal frequency and damping respectively. We recall that negative real parts assures the stability.

Consider that a linear feedback control law like the one in (69) is applied to the system. Substitution of the control law into the system equation (57) leads to

$$\dot{x} = (F + GD)x \tag{72}$$

This equation represents the closed loop control with a system matrix $F + GD$ and its eigenvalues determine the stability and the dynamics of the controlled system. From these eigenvalues we can interpret the effect of the control in terms of changes in the modal frequencies and dampings.

We can use this fact not only for the analysis of the effect of the control for a given feedback strategy, but also as a base for the design of a specific gain matrix. The so–called *pole placement* methods essentially consist of choosing a gain matrix D in such a way that the eigenvalues of the closed loop system matrix take values prescribed a priori by the designer.

3.4. Application example

To illustrate the use of the optimal control law described above, we consider here its application to one of the SUNY–Buffalo experimental structural systems with active tendons, whose diagram is shown in Figure 10.

With the rigid braces on upper floors, the structure can be modeled as a SDOF whose horizontal motion is described by

$$\ddot{d}(t) + 2\nu\omega_o\dot{d}(t) + \omega_o^2 d(t) = -a(t) - \frac{4k_c \cos\alpha}{m}u_o(t) \tag{73}$$

where d is the horizontal relative displacement of the first floor, a the base acceleration and u_o the actuator's piston/rod displacement. m, ν and ω_o are, respectively, the mass, damping and undamped natural frequency. k_c is the stiffness of the cables and α the angle they have with the horizontal. These parameters are:

$$m = 2922.7\,Kg \qquad \nu = 1.24\% \qquad \omega_o = 21.79\,rad/s$$

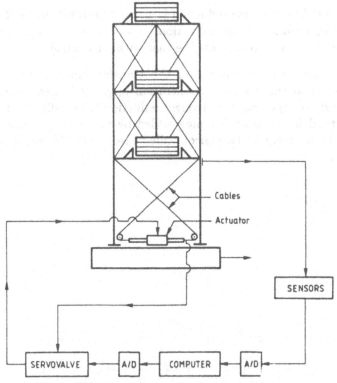

Fig.10.–SUNY–Buffalo experimental control system.

$$k_c = 371950.8 \, N/m \qquad \alpha = 36°$$

By defining the state vector $\mathbf{x}^T = (d, \dot{d})$, eq. (73) can be written in the state form

$$\dot{\mathbf{x}} = \mathbf{F}\mathbf{x} + \mathbf{G}u_o + \mathbf{w} \tag{74}$$

where

$$\mathbf{F} = \begin{pmatrix} 0 & 1 \\ -\omega_o^2 & -2\nu\omega_o \end{pmatrix} \qquad \mathbf{G} = \begin{pmatrix} 0 \\ \frac{-4k_c \cos \alpha}{m} \end{pmatrix} \qquad \mathbf{w} = \begin{pmatrix} 0 \\ -a \end{pmatrix} \tag{75}$$

Eq.74 represents the controlled movement of the structure.

The control problem now consists in generating the actuator displacement $u_o(t)$ such that the performance index defined in (58) is minimized. Consider

$$S = 0; \qquad Q = \begin{pmatrix} k & 0 \\ 0 & 0 \end{pmatrix}; \qquad R = \beta k_c$$

where k (structural stiffness)$= m\omega_o^2 = 1387710$ N/m.

β is a weighting parameter balancing the relative importance of the response reduction (control effectiveness) and the control action requirement (economy).

$\beta < 1$ means that the response reduction is more important, while $\beta > 1$ imposes the desire in keeping a small control action. $\beta = 1$ gives the same importance to both requirements. $\beta = \infty$ represents the case without control.

Figure 11 shows the magnitude of the transfer function between the base acceleration $a(t)$ and the response acceleration $\ddot{d}(t)$. It is clear the reduction of the response and, as expected, the stronger reduction for smaller β. This reduction can be understood in terms of damping added by the control action. For $\beta = 5$ damping is increased from 1.24% (uncontrolled case) to 17.8%, while is 34% for the case with $\beta = 1$.

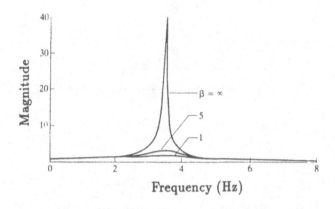

Fig.11.–Transfer function magnitude (from the book "Active Structural Control" by T.T. Soong).

4. PREDICTIVE CONTROL

4.1. Strategy

Predictive control is presented here as representative of digital control methodologies. It is developed within the framework of Figure 3, where the problem of structural control is formulated in discrete–time with the aim of deriving a control algorithm.

From an intuitive point of view, the idea of this method is to use a discrete–time model of the process to predict, at each sampling instant, the response over some interval based on the information available at this instant and as a function of a future control sequence to be determined. Figure 12 illustrates that, from information on the known inputs and states of the process at instant k, we can consider a future

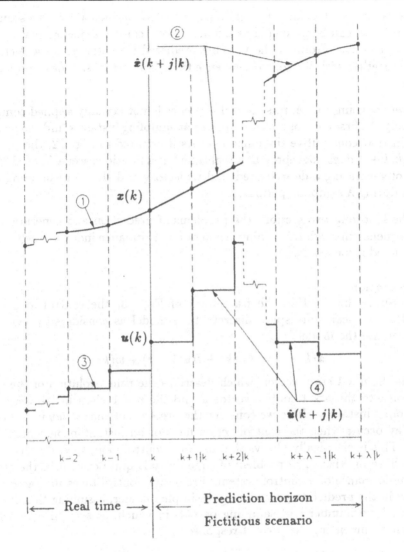

$$\dot{x}(k+j|k)$$

$$x(k)$$

$$u(k)$$

$$\dot{u}(k+j|k)$$

k − 2   k − 1   k   k + 1|k   k + 2|k   k + λ − 1|k   k + λ|k

|← Real time →|← Prediction horizon →|
Fictitious scenario

Present instant

① State trajectory
② Predicted state trajectory
③ Applied control action
④ Control sequence on the prediction scenario

Fig. 12.–Prediction horizon for predictive control.

time interval $[k, k+\lambda]$, defined by a number of control periods λ, where a sequence of states $\hat{x}(k+j|k)$ can be predicted as a function of a control sequence $\hat{u}(k+j-1|k)$, where $j = 1, \cdots, \lambda$. From all the possible predicted trajectories, we will chose the one that, together with the control sequence that produces it, satisfies a performance criterion.

As we are going to see, this control sequence is not actually applied completely. In fact, only the first control vector is applied at sampling instant k and the procedure is re–defined at consecutive time instants. As illustrated in Fig. 12, the prediction horizon $[k, k+\lambda]$ does not represent a real time but a fictitious scenario used with the purpose of generating a desired predicted trajectory and the corresponding control sequence during λ sampling periods.

In the following two sections the problems of prediction and generation of the control sequence through the minimization of a performance index at each instant k are considered respectively.

4.2. Prediction
We assume that, within the framework of Fig. 3, the control loop can be described by a linear state space discrete–time model as considered previously in this chapter. Be the model

$$x(k+1) = Ax(k) + Bu(k-r) + w(k) \tag{76}$$

where x is the $n \times 1$ state vector, which describes the time evolution of the system as a response to the $p \times 1$ control vector u and the $n \times 1$ disturbance vector w at each sampling instant k. Here we consider the presence of *time delay* in the control loop. Delay occurs when the control action has not an instantaneous effect on the response. This is practically always the case in a control loop, specially due to that actuators have inertias. The problem of time delay is quite critical in the stability and the performance of a control system. Predictive control takes into account the time delay in the prediction procedure in a simple manner, assuming that it can be represented by a number r of sampling periods in which, as seen in eq. (76), the control is not influencing the present response.

We can take eq. (76) as source to define a state predictive model in the form

$$\hat{x}(k+j|k) = \hat{A}\hat{x}(k+j-1|k) + \hat{B}\hat{u}(k+j-1-\hat{r}|k)$$
$$(j = 1, 2, \cdots, l+\hat{r}) \tag{77}$$

where $\hat{x}(k+j|k)$ denotes the state vector predicted at instant k for instant $k+j$ and $\hat{u}(\cdot|k)$ denotes the sequence of control vectors on the prediction interval.

This model is redefined at each sampling instant k from the actual state vector and the controls previously applied, i.e.,

$$\hat{x}(k|k) = x(k); \qquad \hat{u}(k-j|k) = u(k-j) \quad (j = 1, \cdots, \hat{r}) \tag{78}$$

Comparing (78) with (77), we do not include the disturbance since it is assumed to be unknown. It would be straightforward to include measurable disturbances in the formulation that we will develop in the following, but, with the purpose of simplifying the presentation, we will not do it. In the predictive model we denote parameters as $\hat{A}, \hat{B}, \hat{r}$ to distinguish to those of the system model (76), assuming that we may use estimates of them.

4.3. Performance criterion and control action

We can consider a linear quadratic performance index of the form

$$
J_k = \frac{1}{2} \sum_{j=1}^{\hat{r}+l} [\hat{x}(k+j|k) - \hat{x}_r(k+j|k)]^T Q_j [\hat{x}(k+j|k) - \hat{x}_r(k+j|k)] +
$$
$$
+ \frac{1}{2} \sum_{j=0}^{l-1} \hat{u}(k+j|k)^T R_j \hat{u}(k+j|k)
$$
(79)

where $\hat{x}_r(k+j|k)$ is a reference trajectory for the state vector, which may be redefined at each sampling instant k from the current state vector and evolve towards the setpoint according to a chosen dynamics. In structural control problems, the setpoint is usually the equilibrium state. Q_j, R_j are symmetric weighting matrices.

By applying (77) recursively from initial conditions (78), we may write

$$
\hat{x}(k+1|k) = \hat{A}x(k) + \hat{B}u(k-\hat{r})
$$
$$
\hat{x}(k+2|k) = \hat{A}^2 x(k) + \hat{A}\hat{B}u(k-\hat{r}) + \hat{B}u(k-\hat{r}+1)
$$
$$
\cdots
$$
$$
\hat{x}(k+\hat{r}|k) = \hat{A}^{\hat{r}} x(k) + \hat{A}^{\hat{r}-1}\hat{B}u(k-\hat{r}) + \cdots + \hat{B}u(k-1)
$$
$$
\hat{x}(k+\hat{r}+1|k) = \hat{A}^{\hat{r}+1} x(k) + \hat{A}^{\hat{r}}\hat{B}u(k-\hat{r}) + \cdots + \hat{A}\hat{B}u(k-1) +
$$
$$
+ \hat{B}\hat{u}(k|k)
$$
(80)
$$
\hat{x}(k+\hat{r}+2|k) = \hat{A}^{\hat{r}+2} x(k) + \hat{A}^{\hat{r}+1}\hat{B}u(k-\hat{r}) + \cdots + \hat{A}^2\hat{B}u(k-1) +
$$
$$
+ \hat{A}\hat{B}\hat{u}(k|k) + \hat{B}\hat{u}(k+1|k)
$$
$$
\cdots
$$
$$
\hat{x}(k+\hat{r}+l|k) = \hat{A}^{\hat{r}+l} x(k) + \hat{A}^{\hat{r}+l-1}\hat{B}u(k-\hat{r}) + \cdots + \hat{A}^l\hat{B}u(k-1) +
$$
$$
+ \hat{A}^{l-1}\hat{B}\hat{u}(k|k) + \hat{A}^{l-2}\hat{B}\hat{u}(k+1|k) + \cdots \hat{B}\hat{u}(k+l-1|k)
$$

Defining the $(\hat{r}+l)n \times 1$ vectors

$$\hat{X} = [\hat{x}(k+1|k)^T, \cdots, \hat{x}(k+\hat{r}|k)^T, \cdots, \hat{x}(k+\hat{r}+l|k)^T]^T$$
$$X_r = [x_r(k+1|k)^T, \cdots, x_r(k+\hat{r}|k)^T, \cdots, x_r(k+\hat{r}+l|k)^T]^T$$

and the $\hat{r}p \times 1$ and $lp \times 1$ vectors

$$U_k = [u(k-1), u(k-2), \cdots u(k-\hat{r})]^T$$
$$\hat{U} = [\hat{u}(k|k), \hat{u}(k+1|k), \cdots, \hat{u}(k+l-1|k)]^T$$

the index (79) can be written in the packed form

$$J_k = \frac{1}{2}[\hat{X} - X_r]^T Q[\hat{X} - X_r] + \frac{1}{2}\hat{U}^T R\hat{U} \tag{81}$$

where the extended weighting matrices Q, R are

$$Q = \mathrm{diag}\,[Q_1, \cdots, Q_{\hat{r}}, \cdots, Q_{\hat{r}+l}]$$
$$R = \mathrm{diag}\,[R_0, \cdots, R_{l-1}]$$

The set of $\hat{r} + l$ equations (80) can be packed in the form

$$\hat{X} = Zx(k) + TU_k + N\hat{U} \tag{82}$$

where Z, T, N are matrices (with dimensions $(\hat{r}+l)n \times n$, $(\hat{r}+l)n \times \hat{r}p$, $(\hat{r}+l)n \times lp$ respectively) defined as

$$Z = \begin{pmatrix} \hat{A} \\ \hat{A}^2 \\ \vdots \\ \hat{A}^{\hat{r}+l} \end{pmatrix} \qquad T = \begin{pmatrix} 0 & 0 & \cdots & 0 & \hat{B} \\ 0 & 0 & \cdots & \hat{B} & \hat{A}\hat{B} \\ \vdots & \vdots & \cdots & \vdots & \vdots \\ \hat{B} & \hat{A}\hat{B} & \cdots & \hat{A}^{\hat{r}-2}\hat{B} & \hat{A}^{\hat{r}-1}\hat{B} \\ \hat{A}\hat{B} & \hat{A}^2\hat{B} & \cdots & \hat{A}^{\hat{r}-1}\hat{B} & \hat{A}^{\hat{r}}\hat{B} \\ \vdots & \vdots & \cdots & \vdots & \vdots \\ \hat{A}^l\hat{B} & \hat{A}^{l-1}\hat{B} & \cdots & \hat{A}^{\hat{r}+l-1}\hat{B} & \hat{A}^{\hat{r}+l}\hat{B} \end{pmatrix}$$

$$N = \begin{pmatrix} 0 & 0 & 0 & \cdots & 0 & 0 \\ \vdots & \vdots & \vdots & \cdots & \vdots & \vdots \\ 0 & 0 & 0 & \cdots & 0 & 0 \\ \hat{B} & 0 & 0 & \cdots & 0 & 0 \\ \hat{A}\hat{B} & \hat{B} & 0 & \cdots & 0 & 0 \\ \vdots & \vdots & \vdots & \cdots & \vdots & \vdots \\ \hat{A}^{l-1}\hat{B} & \hat{A}^{l-2}\hat{B} & \hat{A}^{l-3}\hat{B} & \cdots & \hat{A}\hat{B} & \hat{B} \end{pmatrix}$$

If we substitute (82) into (81), the performance index J_k will be a scalar function of the unknown vector $\hat{U}$. A necessary condition for J_k to be minimum requires its gradient to be zero. Then, substituting (82) into (81) and applying

$$\frac{\partial J_k}{\partial \hat{U}} = 0$$

we obtain

$$\hat{U} = -LZx(k) - LTU_k + LX_r$$

where

$$L = (N^T Q N + R)^{-1} N^T Q$$

The control vector $u(k)$ applied to the process at time k is the first vector component of $\hat{U}$, i.e.,

$$u(k) = \hat{u}(k|k) = -D_1 x(k) - D_2 U_k + D_3 X_r \tag{83}$$

D_1 is a $p \times n$ feedback state gain matrix made up with the first p rows of matrix LZ. D_2 is a $p \times p$ matrix including the first p rows of matrix LT and D_3 is a $p \times n$ matrix having the first p rows of L.

The implementation of the control law (83) can be computationally expensive, particularly if we are dealing with high dimensional systems. Then, in an attempt to reduce the calculations involved in the minimization of the index (79), we can introduce an additional condition consisting of imposing a specified shape to the control sequence. Particular choices of this shape may be that of a step or a pulse. In the case of a step–shaped control sequence, we have a control constant over the prediction interval. That is

$$\hat{u}(k|k) = \hat{u}(k+1|k) = \cdots = \hat{u}(k+l-1|k) \tag{84}$$

Also we can use a simplified version of index (79) in the form

$$J_k = \frac{1}{2}[\hat{x}(k+\hat{r}+l|k) - x_r(k+\hat{r}+l|k)]^T Q'$$

$$[\hat{x}(k+\hat{r}+l|k) - x_r(k+\hat{r}+l|k)] + \frac{1}{2}\hat{u}(k|k)^T R'\hat{u}(k|k) \tag{85}$$

Using (84) in the last prediction of (80), we may write

$$\hat{x}(k+\hat{r}+l|k) = Z^* x(k) + T^* U_k + N^* \hat{u}(k|k) \tag{86}$$

where

$$Z^* = \hat{A}^{\hat{r}+l}$$

$$T^* = [\hat{A}^l \hat{B}\ \hat{A}^{l-1}\hat{B} \cdots \hat{A}^{\hat{r}+l-1}\hat{B}\ \hat{A}^{\hat{r}+l}\hat{B}] \tag{87}$$

$$N^* = \hat{A}^{l-1}\hat{B} + \hat{A}^{l-2}\hat{B} + \cdots + \hat{A}\hat{B} + \hat{B}$$

Substituting (86) into (85) and applying

$$\frac{\partial J_k}{\partial \hat{u}(k|k)} = 0$$

we obtain the control law

$$u(k) = \hat{u}(k|k) = -D_1^* x(k) - D_2^* U_k + D_3^* x_r(k+\hat{r}+l|k) \tag{88}$$

where

$$D_1^* = D_3^* Z^*$$

$$D_2^* = D_3^* T^* \tag{89}$$

$$D_3^* = (N^{*^T} Q' N^* + R')^{-1} N^{*^T} Q'$$

This control law is the most frequently used in the applications of predictive control in the context of active vibration control of structures. For the implementation, the designer has to choose the parameters $\lambda, \hat{r}$ related to the prediction horizon and weighting matrices Q', R', which have to be positive semidefinite and R' non singular. Also matrices $\hat{A}, \hat{B}$ of the predictive model are required. They can be obtained starting from the model of the structure in state space and going through the discretization procedure for the state equation. With all these parameters, matrices D_1^*, D_2^*, D_3^* are computed prior to the implementation using (87) and (89). Note that D_1^* is a gain matrix, while D_2^* contains "memory" matrices multiplying previous controls to compensate for the time delays.

4.4. Application example

To illustrate the use of the predictive control law (83), we go back to the experimental scheme described in Figure 11. Now the digital nature of the closed loop is taken into account to formulate a discrete–time state equation like the one considered in (76). Also we include in our description the existence of a time delay due to the inertia of the active tendon system. This time delay can be modelled considering that the real actuator displacement u_o is related to the control signal given by the controller in the form

$$u_o(t) = u(t - \tau)$$

τ being this time delay. Applying the discretization procedure (previously outlined in this chapter) to the continuous time eq. (74), we can simulate the operation of the closed loop with a sampling period T, obtaining

$$x(k+1) = Ax(k) + Bu(k-d) + Pw(k) \tag{90}$$

where
$$A = \exp(TF); \qquad P = F^{-1}(A - I); \qquad B = PG \tag{91}$$

d is an integer defined as $d = \tau/T$ what implies to assume that the time delay τ can be measured as an integer multiple of the sampling period. This is the most common assumption about time delay in digital control implementations, but eq. (90) can be easily modified for no integer time delays.

4.5. Experimental results and discussion

The model in eq.(90) describes the closed loop basic operations of Figure 11 at each sampling instant and it is used as base model for the formulation of the predictive control law (83) with the following parameters being chosen for the calculation of matrices in (87)–(89): sampling period T=0.01 s, number of delays d=2, reference trajectory x_r=0, matrices $\hat{A}, \hat{B}$ as given in (91). The weighting matrix Q has been chosen as

$$Q = \begin{pmatrix} 1 & 0 \\ 0 & 0 \end{pmatrix}$$

while R is now an scalar defined as $R = rk_c$, r being the weighting factor on the control action.

Results of two series of experiments using the SUNY–Buffalo shaking table are described now in which the base of the structure has been subjected to a band-limited white noise and to an earthquake acceleration respectively. Different tests have been distinguished by the values of the prediction horizon parameter λ and the weighting factor r assigned to the control algorithm.

White noise excitation experiments

The shaking table has been programmed to supply a 0–8 Hz banded white noise acceleration to the base of the structure along about three minutes. The amplitude of the transfer function between the first floor acceleration and the base acceleration has been obtained by means of a spectrum analyzer and used to illustrate the results of the different control tests. Table 1 gives the peak value of the transfer function and its corresponding frequency for the uncontrolled structure as well as the structure under predictive control with different parameters λ and r.

TABLE 3:   White noise excitation experiments

λ	r	Transfer function peak	Frequency (Hz)
No control	No control	154.00	3.52
7	6	9.99	3.60
7	4	8.35	3.60
7	2	4.63	3.68
7	0	2.65	3.76
8	0	3.18	3.68
9	0	4.21	3.68

Figure 13 shows the plots of the magnitude of the transfer function between the response and base accelerations for the controlled cases.

The influence of λ and r on the control can be taken out from these tests. By comparing the transfer functions for λ=7,8 and 9 (with r=0), more reduction in the response is observed for smaller values of λ. By comparing the cases for λ fixed equal to 7 and different values of r=0,2,4 and 6, more reduction is observed as smaller values r are used. These observations are in concordance with the meaning of λ and r. In fact, predictive control law has been derived from the minimization of the performance index in eq. (85). This minimization requires the response predicted for instant $k + \lambda + d$ to be near to the equilibrium state. Thus, the smaller the value of λ is, a more drastic control is imposed with more reduction in the response. Weighting in the control action in the performance index in eq. (85) is imposed by $\mathbf{R} = rk_c$. Consequently, for a fixed prediction horizon, a smaller control action is generated increasing the value of r. In all cases a slight increasing in the effective frequency of the controlled system is observed with respect to the frequency of the structure model as shown in Table 1. The damping effect of the control is noticeable in all the cases. In fact, the peak amplitude of the transfer function is reduced from 154 for the uncontrolled case to 9.99 for the control case with $\lambda = 7$ and $r = 6$.

Earthquake excitation experiments

The shaking table has been shaked using the El Centro 1940 N–S accelerogram (scaled to 25% of its maximum intensity) to excite the structure. Figure 14 shows the first 20 s of the measured time histories of the first floor relative displacement and absolute acceleration for the structure without control.

Figure 15 shows the same histories, as well as the control force supplied by the cables, for one of the tests performed under predictive control. For this test, values of $\lambda = 9$ and $r = 0$ have been assigned. By comparing Figures 14 and 15, a significant reduction in the controlled response is observed with a control effort compatible with the actuator limits. This reduction is characterized by the decreasing in the maximum response as well as a damping effect introduced by the controller.

5. SOME PRACTICAL CONSIDERATIONS ON ROBUSTNESS

5.1. Introduction

As we have been doing previously, the design of a state feedback control law essentially relies on the availability of a model of the control loop. Modelling always involves some degree of idealization and mismatches between the real world and the final mathematical model should be expected. These mismatches can be associated to different sources: lack of linearity, errors in estimating the parameters, unknown time variance of some characteristics due to aging or other reasons, unmodelled interactions between the structure and the actuators, no compensation for time delays, etc.

Within this situation the question of *robustness* arises in the sense of the preservation of the performance, and primarily the stability of the closed loop, in the presence of discrepancies (uncertainties) between the actual process to be controlled and the model used to derive the control law.

The problem of control of systems in the presence of uncertainties has been a topic of strong research effort in the last few years. It is out of our scope here to review this field. We will try to point out some aspects within the particular method of *predictive control* described previously. We will concentrate on the efficiency of this method when it is applied having errors in the structural parameters and in the time delay. The efficiency is evaluated in terms of stability and performance using simple and easy to use tools.

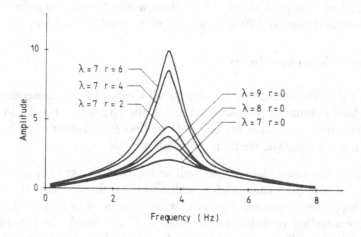

Fig. 13.–Transfer function magnitude with predictive control.

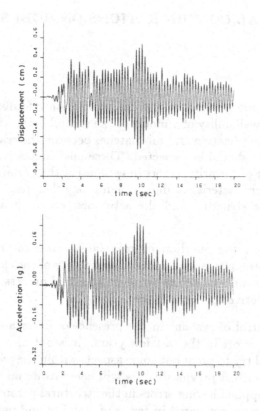

Fig. 14.–Displacement and acceleration responses without control.

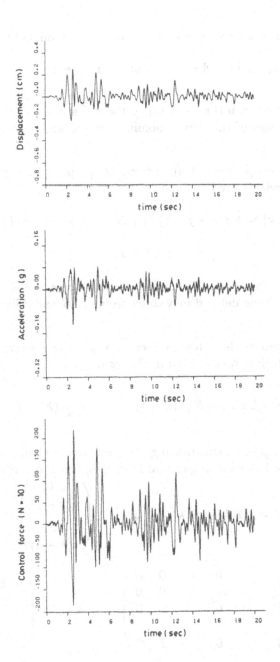

Fig. 15.–Displacement and acceleration responses under predictive control with $\lambda = 9$ and $r = 0$.

5.2. Stability

Consider a process described by a discrete-time state model of the form

$$x(k+1) = A\,x(k) + B\,u(k-d) + w(k) \tag{92}$$

where x is the n state vector, u the r control vector and w the n disturbance vector. A and B are the matrices of the "*true*" nominal system and d is the "*true*" time delay.

Consider the interval $[k, k+\lambda+\hat{d}]$, defined at sampling instant k, where the following predictive model is used

$$\hat{x}(k+j\,|\,k) = \hat{A}\,\hat{x}(k+j-1\,|\,k) + \hat{B}\,\hat{u}(k+j-1-\hat{d}\,|\,k) \tag{93}$$

$$(j = 1,\ldots,\lambda+\hat{d}))$$

This predictive model has a structure similar to the process model 1 but matrices $\hat{A}$ and $\hat{B}$ as well as the time delay $\hat{d}$ may differ from A, B and d due to modelling errors.

Following the procedure described in Section 4.3, we can derive a control law like the one in (80), which now is written in the form

$$u(k) = -\hat{D}x(k) - \hat{K}_1 u(k-1) - \cdots - \hat{K}_{\hat{d}}u(k-\hat{d}) \tag{94}$$

In the case of $\hat{d} \le d \neq 0$, by substituting the predictive control law (94) into the process model (92), the following augmented matrix equation describing the closed loop can be written:

$$\bar{x}(k+1) = \bar{A}\bar{x}(k) + \bar{w}(k); \qquad \bar{x}(k) = \begin{pmatrix} x(k) \\ u(k-1) \\ \vdots \\ u(k-\hat{d}) \end{pmatrix} \tag{95}$$

$$\bar{A} = \begin{pmatrix} A & 0 & \cdots & 0 & \cdots & 0 & B \\ \hat{D} & \hat{K}_1 & \cdots & \hat{K}_{\hat{d}} & \cdots & 0 & 0 \\ 0 & I & \cdots & 0 & \cdots & 0 & 0 \\ & & & \cdots & & & \\ 0 & 0 & \cdots & 0 & \cdots & I & 0 \end{pmatrix} ; \qquad \bar{w}(k) = \begin{pmatrix} w(k) \\ 0 \\ \vdots \\ 0 \end{pmatrix} \tag{96}$$

For other cases different from $\hat{d} \le d \neq 0$, a similar formulation can be considered.

As we know, the closed loop defined by (95) is stable if the eigenvalues of matrix $\bar{A}$ are inside the unit circle. In this case the closed loop is asymptotically and globally stable if the disturbance is $\bar{w}(k) = 0$. If $\bar{w}(k) \neq 0$ is bounded, the augmented state vector is also bounded. The following theorem (given without proof) states the influence of the parameter λ in the stability of the control loop.

Theorem: If the process model (92) and the predictive model (93) are open-loop stable, a value of the prediction horizon λ_0 exists such as for all $\lambda \geq \lambda_0$ the closed loop described by (95) is stable.

5.3. Performance

As a measure of performance, we propose here a set of dimensionless indices that can be evaluated for each mode of vibration. Before defining those indices, we formulate the structural motion in modal coordinates. To do this, consider the following eigenvalue problem associated to the equation of motion (13):

$$(K - \omega^2 M)\, \phi = 0 \tag{97}$$

Eq. (97) is verified for n independent modal vectors $\phi_1, \ldots, \phi_n$ and for n natural frequencies $\omega_1, \ldots, \omega_n$. Uncontrolled mode shapes $\phi_1, \ldots, \phi_n$ are arranged as the columns of Φ, referred as uncontrolled modal matrix.

Eq. (13) can be formulated in modal coordinates by premultiplying it by Φ^T, what results in

$$M^*\, \ddot{\eta}(t) + C^*\, \dot{\eta}(t) + K^*\, \eta(t) = f^*(t) + u^*(t) \tag{98}$$

where M^*, C^* and K^* are the mass, damping and stiffness matrices in modal coordinates given by

$$M^* = \Phi^T M \Phi \qquad C^* = \Phi^T C \Phi \qquad K^* = \Phi^T K \Phi \tag{99}$$

In (98), modal coordinates η verify

$$d(t) = \Phi\, \eta(t) = \phi_1\, \eta_1(t) + \cdots + \phi_n\, \eta_n(t) \tag{100}$$

The modal excitation and control forces f^* and u^* are

$$f^*(t) = \Phi^T f(t); \qquad u^*(t) = \Phi^T u(t) \tag{101}$$

M^* and K^* are diagonal and, if the system is classically damped, C^* is diagonal as well and thus (98) can be decomposed into the set of n scalar equations:

$$\ddot{\eta}_i(t) + 2\,\xi_i\, \omega_i\, \dot{\eta}_i(t) + \omega_i^2\, \eta_i(t) = \frac{f_i^*(t)}{m_i^*} + \frac{u_i^*(t)}{m_i^*} \qquad (i = 1, \ldots, n) \tag{102}$$

where m_i^* is the i-th diagonal element of matrix M^* and ω_i and ξ_i are, respectively, the natural frequency and damping factor of mode i.

These equations can serve as a base for the control problem within the framework of *modal control*. In this case, the problem consists of computing the values of control forces u_i^*. In particular, within the so–called *independent modal space control (IMSC)*, each u_i^* is calculated independently as a feedback of displacement η_i and velocity $\dot{\eta}_i$ and, consequently, the n equations in (102) are uncoupled. Also these modal equations can be used to analyze the effectivenes of a given control law in terms of the effect on each particular mode. It is clear from eq. (100) that the structural displacements d_i are reduced if η_i are reduced. Consequently, the efficiency of the control action can be assessed checking the performance of each individual modal control law.

Consider the operation of a control system over some period of time. We can define the following four dimensionless performance indices $\Delta_{i_1}, \ldots, \Delta_{i_4}$ to evaluate the efficiency of the control of mode i:

$$\Delta_{i_1} = \frac{\max |\eta_i|_{controlled}}{\max |\eta_i|_{uncontrolled}} \qquad \Delta_{i_2} = \frac{\max |\ddot{\eta}_i|_{controlled}}{\max |\ddot{\eta}_i|_{uncontrolled}} \qquad (103)$$

$$\Delta_{i_3} = \frac{\max |u_i^*|}{\max |f_i^*|} \qquad \Delta_{i_4} = \frac{\max |u_i^* \dot{u}_i^*|}{\max |f_i^* \dot{f}_i^*|} \qquad (104)$$

Indices Δ_{i_1} and Δ_{i_2} quantify the reduction of maximum modal displacements and accelerations, respectively. Indices Δ_{i_3} and Δ_{i_4} deal with the maximum required control force and instantaneous power, respectively. Small values of performance indices mean proper control actions but usually it is not possible to reduce all of them: small values of Δ_{i_1} and Δ_{i_2} indicate strong control actions and, consequently, need big values of Δ_{i_3} and Δ_{i_4}. Inversely, smaller Δ_{i_3} and Δ_{i_4} correspond to weaker control actions which generate bigger Δ_{i_1} and Δ_{i_2}. Hence the measure of the quality of the control can be a compromise between such incompatible goals. The relative importance among the four performance indices depends on the design requirements.

One way to use the performance indices described above to assess the efficiency of the control law is to run simulations in which the structure is subject to harmonic excitations. Then the indices are plotted against the period of the excitation, thus providing a sort of *controlled response spectra*.

As an example, we present some of these spectra obtained applying the predictive control law (94) for a single mode structure. This can be characterized by the natural period $T_i = 2\pi/\omega_i$ and damping factor ξ_i. Since the efficiency of the control action appears not to be very sensitive to damping factor, we take $\xi_i = 0$ in most of the cases. The natural period T_i and the period T_0 of the harmonic excitation are divided by the sampling period T of the digital control implementation for normalization purposes. Figures 16 and 17 give controlled response spectra for indices Δ_{i1} and Δ_{i3} respectively for a mode with $T_i/T = 100$. They have been computed for fixed weighting factors and for different values of the horizon prediction length λ, which is the main design parameters. The control law has been derived under *ideal* conditions, with no errors in the parameters and time delay $d = \hat{d} = 0$.

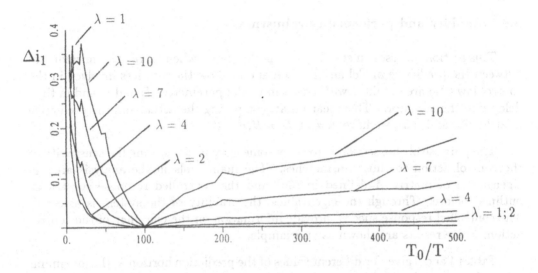

Fig.16.– Controlled response spectrum for index Δ_{i1}.

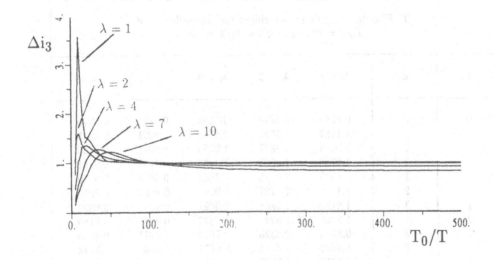

Fig.17.– Controlled response spectrum for index Δ_{i3}.

The minimum values for $T_0/T = 100$ in Figure 16 correspond to the resonant period of the uncontrolled structure and the maxima correspond to the different resonant periods of the controlled cases. Figure 16 shows that smaller values of λ provide smaller values of Δ_{i1}. It confirms a feature that we have seen previously: smaller λ implies more reduction of the response. Figure 17 shows that smaller λ produces bigger maximum action and smaller resonant periods, what implies bigger increase of the resonant frequency and, consequently, stronger control actions.

5.4. Stability and performance robustness

This section focusses on stability and performance when there exist mismatches between the predictive model and the actual one. Now the matrices involved in the control law (94) are obtained with errors in modal parameters T_i and ξ_i and in time delay d in the actuators. This means that, comparing the actual and the predictive models (92) and (93), we have $\hat{A} \neq \hat{A}, \hat{B} \neq \hat{B}, \hat{d} \neq d$.

The purpose of this section is to show some examples assessing the sensitivity of the control action to those mismatches. The main tools in the analysis are the eigenvalues of matrix $\bar{A}$ defined in (96) and the controlled response spectra as outlined before. Through the eigenvalues, the stability of the system is assessed. The controlled response spectra provide information on the efficiency of the control action. Some results are shown as an example.

Tables 4 and 5 give, for different values of the prediction horizon λ, the maximum value of the modulus of the eigenvalues of matrix $\bar{A}$ for discrepancies between d and $\hat{d}$ (Table 4) and for additive identification errors δT_i and $\delta \xi_i$ (Table 5).

		TABLE 4: Maximum modulus of the eigenvalues of $\bar{A}$ $T_i/T = 100; \xi_i = 0; 5 < T_0/T < 500$				
$\hat{d}$	d	$\lambda_i = 1$	$\lambda_i = 2$	$\lambda_i = 4$	$\lambda_i = 7$	$\lambda_i = 10$
0	0	0.8945	0.6328	0.7553	0.8811	0.9070
0	1	1.1487	1.1321	0.8385	0.8651	0.8945
0	2	1.2544	1.2847	1.0553	0.8911	0.8766
0	3	1.2768	1.3066	1.1269	0.9715	0.9161
1	1	0.8945	0.6328	0.7553	0.8811	0.9070
1	2	1.1444	1.3727	0.9299	0.8697	0.8919
1	3	1.2106	1.1861	1.0497	0.9349	0.8954
2	1	1.2909	1.5224	0.9827	0.8968	0.9180
2	2	0.8945	0.6328	0.7553	0.8811	0.9070
2	3	1.1597	1.3515	0.9874	0.8900	0.8914
3	1	1.6317	1.7523	1.2158	0.9103	0.9274
3	2	1.3354	1.4489	1.0343	0.8989	0.9193
3	3	0.8945	0.6328	0.7553	0.8811	0.9070

TABLE 5:   Maximum Modulus of the Eigenvalues of $\bar{A}$
$T_i/T = 20$; $\xi_i = 0.05$; $\hat{d} = d = 1$; $5 < T_0/T < 500$

δT_i [%]	$\delta \xi_i$ [%]	$\lambda_i = 1$	$\lambda_i = 2$	$\lambda_i = 4$	$\lambda_i = 7$	$\lambda_i = 10$
50	0	0.897	0.660	0.767	0.980	0.979
20	0	0.891	0.646	0.765	0.981	0.982
10	0	0.888	0.637	0.764	0.982	0.983
5	0	0.886	0.632	0.764	0.982	0.984
0	0	0.883	0.626	0.763	0.983	0.984
-5	0	0.880	0.618	0.763	0.983	0.984
-10	0	0.877	0.609	0.763	0.983	0.985
-20	0	0.868	0.586	0.764	0.984	0.985
-50	0	0.801	0.927	1.200	0.985	0.984
0	20	0.886	0.632	0.766	0.983	0.984
0	10	0.884	0.629	0.765	0.983	0.984
0	5	0.884	0.627	0.764	0.983	0.984
0	2	0.884	0.626	0.764	0.983	0.984
0	-2	0.883	0.625	0.763	0.983	0.984
0	-5	0.883	0.624	0.763	0.983	0.984

Results in Table 4 illustrate that, even in the presence of severe disagreements between d and $\hat{d}$, a value λ_0 can be found such as the eigenvalues of $\bar{A}$ are inside the unit circle for $\lambda > \lambda_0$, thus preserving the stability of the controlled system. It is worth noting that, provided $d = \hat{d}$, values in Table 4 do not depend on d. Also it is interesting to note that a value of $\lambda = 1$ is enough in order to have stable control when there are no discrepancies between d and $\hat{d}$; $\lambda > 1$ is required when $d \neq \hat{d}$. Results from Table 5 show that errors less than 20% do not significantly modify stability conditions. However, error $\delta T_i = -50\%$ can generate unstable control actions.

In Figure 18 spectra of Δ_{i1} are presented for some cases included in Table 4 with $\lambda = 7$. From Figure 18, it is apparent that the effect of the error in the estimation of the delays is solely important for high frequency excitations.

In Figure 19 spectra of Δ_{i_2} are presented for some cases included in Table 5, with $d = \hat{d} = 1$, $\xi_i = 0.05$ and $\lambda_i = 2$. Plots from Figure 19 show that positive errors generate bigger values of Δ_{i2} while negative errors produce smaller values of such index.

As Tables 4 and 5 show that control action is stable, Figures 17–19 show that it is also efficient.

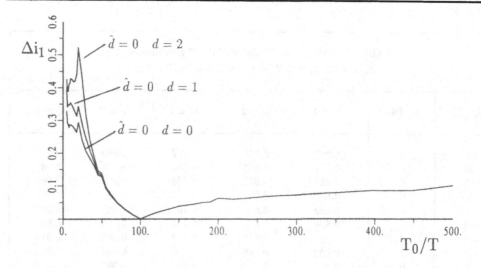

Fig. 18.– Controlled response spectra for index Δ_{i1} with errors in the time delay.

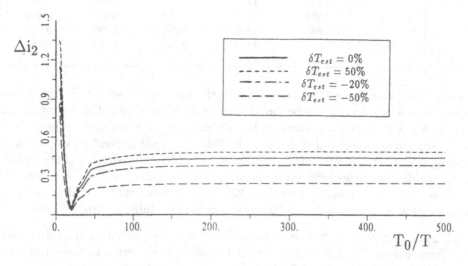

Fig. 19.– Controlled response spectra for index Δ_{i2} with errors in the natural period.

6. BIBLIOGRAPHY

There is an extensive literature covering the topics outlined in this chapter and many more that have not been included. Here is a short list of some books and papers that can be useful for having a wider and deeper view of problems related to development of algorithms for active structural control in civil engineering. They are classified according the topics they cover.

Control (general)

- Brogan, W.L. (1985). *"Modern Control Theory"*. Prentice–Hall, USA.

- Franklin, G.F., Powell, J.D. (1990). *"Digital Control of Dynamic Systems"*. Addison–Wesley Publishing Company, USA.

- Kuo, B.C. (1991). *"Automatic Control Systems, 6th edition"*. Prentice–Hall, USA.

- Kuo, B.C. (1992). *"Digital Control Systems, 2nd edition"*. Saunders College Publ., USA.

State space representation

- Chen, CT. (1984). *"Linear Systems Theory and Design"*. Holt, Rinehart and Winston, USA.

- Ogata, K. (1967). *"State Space Analysis of Control Systems"*. Prentice–Hall, USA.

- Moler, C., Van Loan, Ch. (1978). *"Nineteen dubious ways to compute the exponential of a matrix"*. SIAM Review, Vol. 20, No.4, pp.801–836.

Active structural control (general)

- Meirovitch, L. (1990). *"Dynamics and Control of Structures"*. John Wiley, USA.

- Soong, T.T. (1990). *"Active Structural Control"*. Longman Scientific & Technical, England.

Optimal control

- Abdel–Rohman, M., Quintana, V.H., Leipholz, H.H.E. (1980). *"Optimal control of civil engineering structures"*. ASCE Journal of Engineering Mechanics, Vol. 106, pp. 57–73.

- Bryson, A.E., Ho, Y. (1969). *"Applied Optimal Control"*. Ginn and Company, USA.

- Kwakernaak H., Sivan, R. (1972). *"Linear Optimal Control Systems"*. Wiley–Interscience, USA.

- Sage, A.P., White, C.C. (1977). *"Optimum Systems Control"*. Prentice–Hall, USA.

- Yang, J.N (1975). *"Application of optimal control theory to civil engineering structures"*. .

Predictive control

- López Almansa, F.,Rodellar, J. (1990). *"Feasibility and robustness of predictive control of building structures"*. Earthquake Engineering and Structural Dynamics, Vol. 19, pp. 157–171.

- Rodellar, J., Barbat, A.H., Martín Sánchez, J.M. (1987). *"Predictive control of structures"*. ASCE Journal of Engineering Mechanics, Vol. 113, pp. 797–812.

- Rodellar, J., Chung, L., Soong, T.T., Reinhorn, A. (1989). *"Experimental digital control of structures"*. ASCE Journal of Engineering Mechanics, Vol. 115, pp. 1245–1261.

Other methods in control of structures

- Abdel–Rohman, M., Leipholz, H.H.E. (1978). *"Structural control by pole assignement method"*. ASCE Journal of Engineering Mechanics, Vol. 104, pp. 1157–1175.

- Kelly, J., Leitmann, G., Soldatos, A (1987). *"Robust control of base isolated structures under earthquake excitation"*. Journal of Optimization Theory and Applications, Vol. 53, pp. 159–181.

- Miller, R.K., Masri, S.F., Dehghanyar, T.J., Caughey, T.K. (1988). *"Active vibration control of large civil structures"*. ASCE Journal of Engineering Mechanics, Vol. 114, pp. 1542–1570.

- Rodellar, J., Leitmann, G., Ryan, E.P. (1993). *"On output feedback control of uncertain coupled systems"*. International Journal of Control, Vol. 58(2), 99. 445–457.

- Yang, J.N, Akbarpour, A., Ghaemmaghami, P. (1987). *"New control algorithms for structural control"*. ASCE Journal of Engineering Mechanics, Vol. 113, pp. 1369–1386.

CHAPTER XV

ACTIVE CONTROL EXPERIMENTS
AND STRUCTURAL TESTING

T.T. Soong
State University of New York at Buffalo, Buffalo, NY, USA

ABSTRACT

As in all other new technological innovations, experimental verification constitutes a crucial element in the maturing process as active structural control progresses from conceptualization to actual implementation. Experimental studies are particularly important in this area since hardware requirements for the fabrication of a feasible active control system for structural applications are in many ways unique. As an example, control of civil engineering structures requires the ability on the part of the control device to generate large control forces with high velocities and fast reaction times. Experimentation on various designs of possible control devices is thus necessary to assess the implementability of theoretical results in the laboratory and in the field.

In order to perform feasibility studies and to carry out control experiments, investigations on active control have focused on several control mechanisms as described below.

ACTIVE BRACING SYSTEM (ABS)

Active control using structural braces and tendons has been one of the most studied mechanisms. Systems of this type generally consist of a set of prestressed tendons or braces connected to a structure whose tensions are controlled by electro-hydraulic servomechanisms. One of the reasons for favoring such a control mechanism has to do with the fact that tendons and braces are already existing members of many structures. Thus, active bracing control can make use of existing structural members and thus minimize extensive additions or modifications of an as-built structure. This is attractive, for example, in the case of retrofitting or strengthening an existing structure.

Active tendon control has been studied analytically in connection with control of slender structures, tall buildings, bridges and offshore structures. Early experiments involving the use of tendons were performed on a series of small-scale structural models (Roorda, 1980), which included a simple cantilever beam, a king-post truss and a free-standing column while control devices varied from tendon control with manual operation to tendon control with servovalve-controlled actuators.

More recently, a comprehensive experimental study was designed and carried out in order to study the feasibility of active bracing control using a series of carefully calibrated structural models. As Fig. 1 shows, the model structures increased in weight and complexity as the experiments progressed from Stage 1 to Stage 3 so that more control features could be incorporated into the experiments. Figure 2 shows a schematic diagram of the model structure studied during the first two stages. It is a three-story steel frame modeling a shear building by the method of mass simulation. At Stage 1, the top two floors were rigidly braced to simulate a single-degree-of-freedom system. The model was mounted on a shaking table which supplied the external load. The control force was transmitted to the structure through two sets of diagonal prestressed tendons mounted on the side frames as indicated in Fig. 2.

Results obtained from this series of experiments are reported in (Chung et al., 1988; Chung et al., 1989). Several significant features of these experiments are noteworthy. First, they were carefully designed in order that a realistic structural control situation could be investigated. Efforts made towards this goal included making the model structure dynamically similar to a real structure, working with a carefully calibrated model, using realistic base excitation, and requiring more realistic control forces. Secondly, these experiments permitted a realistic comparison between analytical and experimental results, which made it possible to perform extrapolation to real structural behavior. Furthermore, important practical considerations such as time delay, robustness of control algorithms, modeling errors and structure-control system interactions could be identified and realistically assessed.

Experimental results show significant reduction of structural motion under the action of the simple tendon system. In the single-degree-of-freedom system case, for example, a reduction of over 50% of the first-floor maximum relative displacement could be achieved. This is due to the fact that the control system was able to induce damping in the system from a damping ratio of 1.24% in the uncontrolled case to 34.0% in the controlled case (Chung et al., 1988).

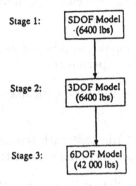

Stage 1: | SDOF Model
            | ·(6400 lbs)

Stage 2: | 3DOF Model
            | (6400 lbs)

Stage 3: | 6DOF Model
            | (42 000 lbs)

Fig. 1   Laboratory Tests of Active Bracing Systems

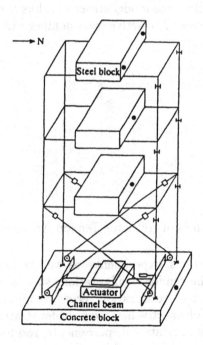

Fig. 2   Schematic Diagram of Model Structure at Stages 1 and 2

As a further step in this direction, a substantially larger and heavier six-story model structure was fabricated for Stage 3 of this experimental undertaking. It is also a welded space frame utilizing artificial mass simulation. It weighs 42,000 lbs and stands 18 ft in height.

Multiple tendon control was possible in this case and the following arrangements were included in this phase of the experiments:

(a) A single actuator is placed at the base with diagonal tendons connected to a single floor.

(b) A single actuator is placed at the base with tendons connected simultaneously to two floors, thus applying proportional control to the structure.

(c) Two actuators are placed at different locations of the structure with two sets of tendons acting independently.

Several typical actuator-tendon arrangements are shown in Fig. 3. Attachment details of the tendon system are similar to those shown in Fig. 2.

Another added feature at this stage was the testing of a second control system, an active mass damper, on the same model structure, thus allowing a performance comparison of these two systems. The active mass damper will be discussed in more detail in the next section.

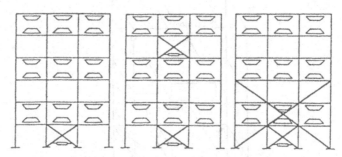

Fig. 3   Examples of Actuator-Tendon Arrangements

For the active tendon systems, experimental as well as simulation results have been obtained based upon the tendon configurations stipulated above. Using the N-S component of the El-Centro acceleration record as input, but scaled to 25% of its actual intensity, control effectiveness was demonstrated. For example, in terms of reduction of maximum relative displacements, results under all actuator-tendon arrangements tended to cluster within a narrow range. At the top floor, a reduction of 45% could be achieved. Control force and power requirements were also found to be well within practical limits when extrapolated to the full-scale situation (Reinhorn et al., 1989).

It is instructive to give more details of the experimental set-up, results obtained and their implications with regard to all the experiments described above. To conserve space, however, this will be done only for the experiments performed at Stage 2.

Stage 2 Experiments

As described above, the basic experimental set-up used in this study consisted of a three-story 1:4 scale frame with one tendon control device implemented to the first floor (Fig. 2). The control was supplied by a servocontrolled hydraulic actuator through a system of tendons.

The state variable measurements were made by means of strain gage bridges installed on the columns just below each floor slab. For each set of the strain gage bridges, the signal from one strain gage bridge was used as the signal of measured storydrift displacement between adjoining stories, while the signal from the second set was further passed through an analog differentiator to yield measured storydrift velocity. The base acceleration and absolute acceleration of each floor were directly measured by the use of accelerometers installed at the base of the structure and on the floor slabs. The transducers and instrumentation system are shown in Fig. 2. A block diagram showing the measurement system and the control procedure is given in Fig. 4.

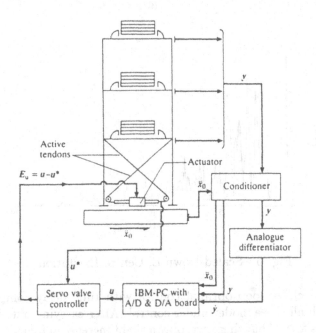

Fig. 4   Block Diagram of Control System

The model was shaken by means of a shaking table with banded white noise and an earthquake accelerogram. Under white noise excitation, modal properties were identified from the frequency response functions for system identification. Moreover, it provided a preliminary examination of the system performance including structural, sensor and controller dynamics for more realistic inputs that were to follow. The N-S component of El-Centro acceleration record was used in the experiment, however, it was scaled to 25% of its actual intensity to prevent inelastic deformations in the model structure during uncontrolled vibrations. The reproduced time history and the frequency distribution of the scaled down El Centro excitation are shown in Fig. 5.

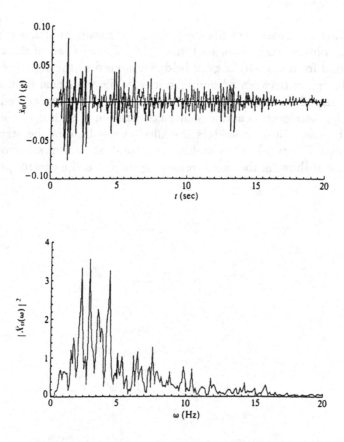

Fig. 5   Scaled-Down El Centro Excitation

The classical closed-loop optimal control with time delay compensation was first studied with all three modes under control. After carrying out the variational procedure, it was found that there was only a slight increase in frequencies (stiffness) but damping factors were increased from 1.62%, 0.3% and 0.36% to 12.77% 12.27% and 5.45% (Tables 1 and 2).

Table 1   Parameters of the Modal Structure

Mass matrix M (lb-sec²/in)	$\begin{bmatrix} 5.6 & 0 & 0 \\ 0 & 5.6 & 0 \\ 0 & 0 & 5.6 \end{bmatrix}$
Stiffness matrix K (lb/in)	$\begin{bmatrix} 15\,649 & -9\,370 & 2107 \\ -9\,370 & 17\,250 & -9274 \\ 2\,107 & -9\,274 & 7612 \end{bmatrix}$
Damping matrix C (lb-sec/in)	$\begin{bmatrix} 2.185 & -0.327 & 0.352 \\ -0.327 & 2.608 & -0.015 \\ 0.352 & -0.015 & 2.497 \end{bmatrix}$
Modal frequency ω (Hz)	$\begin{bmatrix} 2.24 \\ 6.83 \\ 11.53 \end{bmatrix}$
Modal damping factor ζ (%)	$\begin{bmatrix} 1.62 \\ 0.39 \\ 0.36 \end{bmatrix}$
Tendon stiffness k_c (lb/in)	2124
Tendon inclination α (°)	36
Modal matrix Φ	$\begin{bmatrix} 0.262 & 0.743 & 0.583 \\ 0.568 & 0.373 & -0.728 \\ 0.780 & -0.555 & 0.360 \end{bmatrix}$

Table 2   Parameters of Control System

Parameters	Three controlled modes	One controlled mode
Response weighting matrix $Q^{[1]}$	$\begin{bmatrix} K & 0 \\ 0 & 0 \end{bmatrix}$	
Control weighting matrix $R^{[2]}$	$20\,k_c$	
Modal frequency ω (Hz)	$\begin{bmatrix} 2.28 \\ 6.94 \\ 11.56 \end{bmatrix}$	$\begin{bmatrix} 2.28 \\ 6.83 \\ 11.53 \end{bmatrix}$
Modal damping factor ζ (%)	$\begin{bmatrix} 12.77 \\ 12.27 \\ 5.45 \end{bmatrix}$	$\begin{bmatrix} 1.62 \\ 0.39 \\ 0.36 \end{bmatrix}$
Time delay $\tau_x, \tau_{\dot{x}}$ (msec)	35	88
Feedback gain matrix G^{T}	$\begin{bmatrix} 0.1857 \\ -0.1571 \\ 0.0641 \\ 0.0171 \\ 0.0021 \\ 0.0055 \end{bmatrix}$	$\begin{bmatrix} 0.0056 \\ 0.0123 \\ 0.0157 \\ 0.0027 \\ 0.0059 \\ 0.0076 \end{bmatrix}$

[1] K is structural stiffness matrix
[2] k_c is tendon stiffness

The spillover was investigated by selecting the first fundamental mode as the controlled critical mode. The critical modal quantities were reconstructed from the measurements at all floors. The effect of spillover to the residual modes was studied. When fewer output measurements were available, the estimated critical modal quantities were actually affected by the observation spillover to the residual modes. Even worse, time delay was compensated as if the outputs were contributed by the critical modes alone. The combined effect of observation spillover and time delay made the system unstable.

When the first fundamental mode was the only controlled critical mode, the modal quantities were recovered from measurements at all three floors. In the presence of modeling errors (mode shapes were not exactly orthogonal) and measurement noise, the first modal quantities could not be reconstructed perfectly and small contribution of the residual modes to the feedback signal was unavoidable. Because of small stability margins (small damping factors) for the second and third modes, the model structure was very sensitive to these errors. To circumvent this problem, the command control signal was passed through a low-pass filter before driving the actuator in order to eliminate effect of the residual modes. However, no perfect filter exists; the higher the order of the filter, the sharper is the cutoff frequency, but the longer is the time delay. As a compromise, a third-order Butterworth filter with a cutoff frequency of 5 Hz was selected, but time delay was increased from 35 msec. to 88 msec.

Acceleration frequency response functions as shown in Figs 6-8 were constructed by using banded white noise excitation. For the three controlled modes, significant damping effect (large active damping) was reflected from a decrease in peak magnitudes, but peak frequencies shifted to the right due to small active stiffness. It was shown that all three modes were under control with one controller in the presence of time delay. For the case of one controlled mode, the peak of the first mode was decreased but the peaks of the second and third modes were increased. Due to the effect of control spillover, the performance of the controlled system was not better than that of the uncontrolled one.

Under El Centro excitation, significant reduction in acceleration was achieved with three controlled modes. In addition to the reduction in peak magnitudes, the effect of active damping was clearly evident due to control execution but the excitation frequency was distributed over all three modes. Due to control spillover, the control effect was greatly degraded (Figs. 9-11).

The instantaneous optimal control algorithms were also studied with all three modes under control using the seismic excitation. With carefully chosen weighting matrices, similar control effects could also be achieved.

Good agreement was achieved between analytical and experimental results. The discrepancies were larger in the uncontrolled test due to the servocontrolled system.

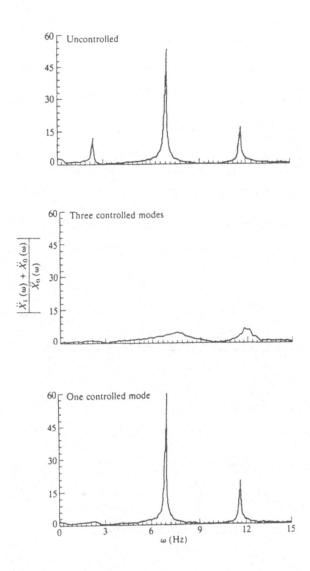

Fig. 6   First-Floor Acceleration Frequency Response Functions

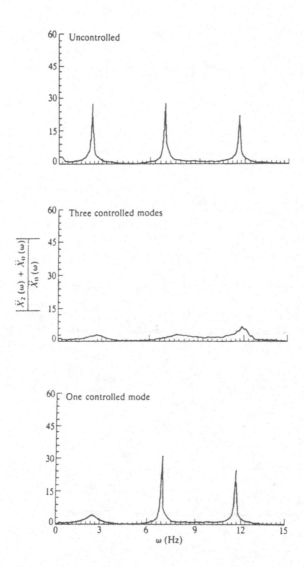

Fig. 7   Second-Floor Acceleration Frequency Response Functions

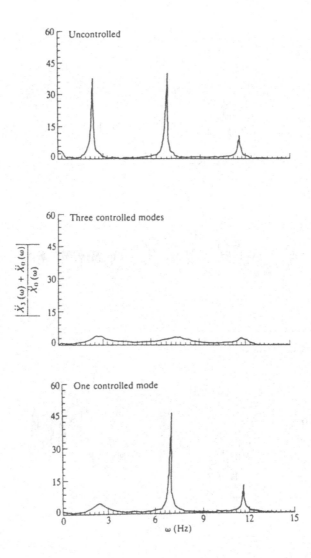

Fig. 8   Third-Floor Acceleration Frequency Response Functions

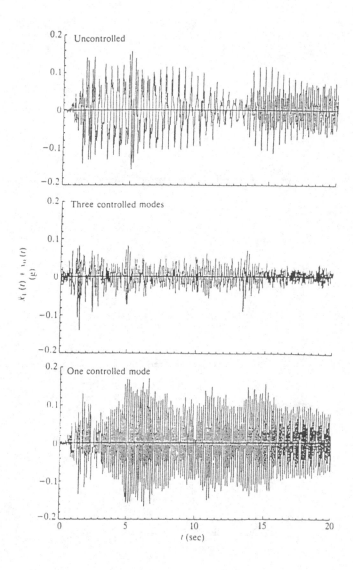

Fig. 9 First-Floor Accelerations Under El Centro Excitation

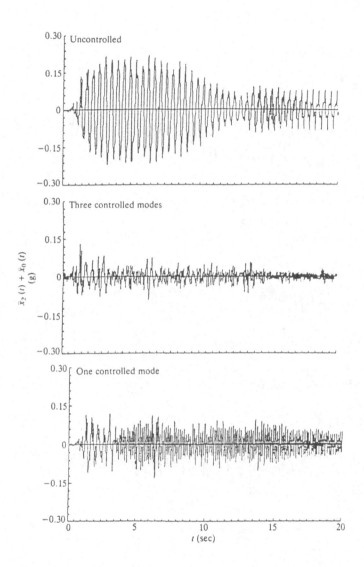

Fig. 10 Second-Floor Accelerations Under El Centro Excitation

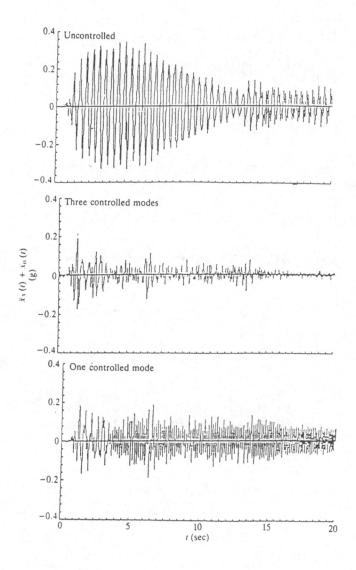

Fig. 11   Third-Floor Accelerations Under El Centro Excitation

The actuator was kept stationary by this system during uncontrolled tests. However, slight actuator movement was induced by the structural motion and the actuator movement was continuously corrected to reduce the error to zero. This interaction between the controller and the structure made the damping force a complicated function of the structural response. For the case of El Centro excitation, some discrepancies resulted from the fact that the equivalent viscous damping was different from the calibrated one measured in the banded white noise tests. However, for the controlled cases, most of the damping was contributed by the feedback force. Therefore, the influence of actuator-structure interaction was negligible and excellent agreement was observed. With one controlled mode, the control force was of a lower magnitude and of a lower frequency, leading to a better performance of the actuator and hence excellent agreement between experimental and analytical results.

The results presented above are encouraging in that they show simple control systems can be effective for response control of complex structures. In addition, extrapolations show that tendon control can be feasible for full-scale structural applications in terms of force and power requirements.

ACTIVE MASS DAMPER AND ACTIVE MASS DRIVER (AMD)

The study of this control mechanism was in part motivated by the fact that passive tuned mass dampers for motion control of tall buildings are already in existence. Tuned mass dampers are in general tuned to the first fundamental frequency of the structure, thus only effective for building control when the first mode is the dominant vibrational mode. This may not be the case; however, when the structure is subjected to seismic forces when vibrational energy is spread over a wider frequency band. It is thus natural to ask what additional benefits can be derived when they function according to active control principles. Indeed, a series of feasibility studies of active and semi-active mass dampers have been made along these lines and they show, as expected, enhanced effectiveness for tall buildings under either strong earthquakes or severe wind loads.

Recently, experimental studies of active mass damper systems have been carried out in the laboratory using scaled-down building models. In the work of Kuroiwa and Aizawa (1987), an AMD was placed on top of a four-story model frame. The moving mass was a variable, ranging from approximately 1% to 2% of the structural weight. The model structure, 1 m (width) × 1 m (depth) × 2 m (height) and weighing 970 kg, was placed on a shaking table which provided simulated earthquake-type base motion. Following a closed-loop control algorithm and using three representative earthquake inputs, experimental results show that the maximum relative displacement reduction at the top floor could be as high as 50%; however, only 5-7% reduction was possible for the maximum absolute acceleration of the top floor.

In another experimental study, a moving mass termed an "active mass driver" was placed on a 0.5 m (width) × 3 m (height) three-story steel frame. The structure was mounted on a shaking table while an electro-magnetic force generator was adopted as the active controller. Experimental results indicate that a two-thirds reduction of the maximum acceleration and displacement could be achieved.

At a much larger scale, an active mass damper system was tested in conjunction with an active tendon system as described above. Using the same six-story 42,000-lb structure, the AMD system was placed on top of the structure, which could be operated under different conditions by changing its added mass, it stiffness and the state of the regulator. A total of 12 cases were performed in the experiment.

Extensive experimental results were obtained under various simulated earthquake excitations. A summary of results obtained under the 25% intensity El Centro excitation is given below:

Percent Reduction of Maximum Top-floor Relative Displacement: 43.3-57.2
Percent Reduction of Maximum Top-floor Acceleration: 5.5-30.7
Percent Reduction of Maximum Base Shear: 31.4-44.4
Maximum Control Force Required (kips):   0.68-2.56
Maximum Mass Peak-to-Peak Stroke (in):   3.23-10.1
Maximum Control Power Required (Kw):   0.82-5.73

One of the advantages of testing two different active systems using the same model structure is that their performance characteristics can be realistically compared. Extensive simulation and experimental results obtained based on the six-story, 42,000-lb model structure show that both AMD and ABS display similar control effectiveness in terms of reduction in maximum top-floor relative displacement, in maximum top-floor absolute acceleration, and its maximum base shear. They also have similar control requirements such as maximum control force and maximum power. Other information which may shed more light on their relative merits but is not considered here includes cost, space utilization, maintenance and other practical observations.

PULSE GENERATOR

Pulse control has also been a subject of experimental study in the laboratory. This control algorithm was tested using a six-story frame weighing approximately 159 kg and measuring six feet in height (Miller et al., 1987; Traina et al., 1988). Figure 12 shows the model structure together with the test apparatus which includes vibration exciter, instrumentation, pneumatic power supply, and the minicomputer used for digital control. As shown in the figure, electrodynamic exciter, sensor, and pneumatic actuators were located at the top of the structure. The actuators consisted of two solenoids which metered the flow of compressed air at 125 psi through eight nozzles, thus generating the required control pulses.

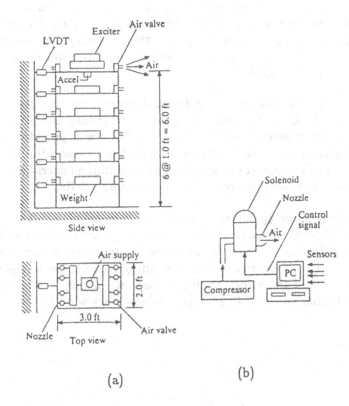

Fig. 12   Pulse Control System
(a) Control Configuration; (b) Pneumatic Control
(Traina et al., 1988)

When the structure was subjected to a harmonic excitation at a frequency close to the fundamental frequency, it was shown that,, within ten periods of onset of control, the response was reduced to approximately 15% of its uncontrolled value.

Discussions on some of the recently developed cold-gas generators having potential structural control applications can be found in (Agababian Assoc., 1984a and 1984b). In addition, pulse control experiments involving hydraulic and electromagnetic actuators have also been conducted in the laboratory (Traina et al., 1988).

AERODYNAMIC APPENDAGE

The use of aerodynamic appendages as active control devices to reduce wind-induced motion of tall buildings has several advantages, its main attractive feature being that the control designer is able to exploit the energy in the wind to control

the structure, which is being excited by the same wind. Thus, it eliminates the
need for an external energy supply to produce the necessary control force; the only
power required is that needed to operate the appendage positioning mechanism.
This type of control is clearly not suitable for seismic applications; however, it is
included here for completeness.

For this control scheme, a wind-tunnel experiment was conducted using an
elastic model at a geometric scale of roughly 1:400 (Soong and Skinner, 1981). This
is schematically presented in Fig. 13. Its stiffness was provided by a steel plate
fixed at the structure core as shown, and its length was adjusted so that under
planned wind conditions in the wind tunnel used in the experiment, the first mode
was dominant and was observed to be approximately 5 Hz.

The aerodynamic appendage consisted of a metal plate. It was controlled by
means of a 24 VDC solenoid, activated by the sign of structural velocity as sensed
by a linear differential transformer, followed by appropriate carrier and signal amplifications and a differentiator. The appendage area normal to the wind direction
was roughly 2% of the structural frontal area when fully extended. A boundary
layer wind tunnel was used to generate the necessary wind forces.

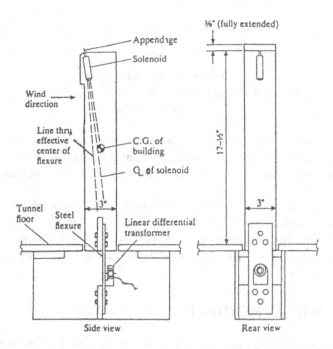

Fig. 13 Schematic Diagram of Aerodynamic Appendage

HYBRID SYSTEMS

Discussed in the above are some of the most studied control mechanisms for structural applications. Many others have been proposed. Furthermore, the combined use of passive devices, such as base isolation, and active devices, such as active tendons or braces, referred to as hybrid systems have been suggested for some specific structural applications (Reinhorn et al., 1987; Kelly et al., 1987; Pu and Kelly, 1991; Inaudi et al., 1992). Hybrid control can alleviate some of the limitations which exist for either the passive system or the active system operating singly, thus leading to a more effective protective system. For example, in combination with a passive system, the force requirement of an active control system can be significantly reduced, which allows the active control device to operate at a much higher efficiency and effectiveness. At the same time, a purely passive system such as a simple elastomeric bearing is limited to low-rise structures because of the possibility of uplift of the isolator due to large horizontal accelerations. The addition of an active system is capable of minimizing this uplift effect.

Recent research on hybrid systems has been focused on combining a base isolation system with an active device. Since base isolation systems exhibit nonlinear behavior, nonlinear and other robust control laws have been developed (Yang et al., 1992a,1992b,1993; Subramaniam et al., 1992; Nagarajaiah et al., 1993; Inaudi et al., 1993) and several small-scale hybrid control experiments have been carried out. In one study (Reinhorn et al., 1987; Wang and Reinhorn, 1989), a sliding isolation system (Fig. 14) was combined with displacement control devices. More recently, a structural model was built and tested with a hybrid control system as shown in Fig. 15 (Riley et al., 1992,1993; Nagarajaiah et al., 1992). The hybrid system consisted of a series of low friction sliding bearings using highly pressurized teflon interface sliding against stainless steel. The system was developed to reduce the absolute acceleration of the foundation using a variety of control algorithms, from variable friction to acceleration feedback.

Other experiments include a hybrid isolation system using friction controllable sliding bearings (Feng et al., 1991). The structural model with the hybrid sliding isolation device is shown in Fig. 16 and the details of the bearing is shown in Fig. 17. During earthquakes, this isolation system controls the friction force on the sliding interface between the support structure and the ground by adjusting the bearing chamber pressure in order to confine the sliding displacement within an acceptable range.

These experiments were successful in demonstrating the advantages of a hybrid control system over a passive base isolation system in motion reduction of the superstructure. Verification of its practical feasibility, however, still awaits results using more realistic structural models or full-scale structures. The cost associated with hybrid systems is another important consideration.

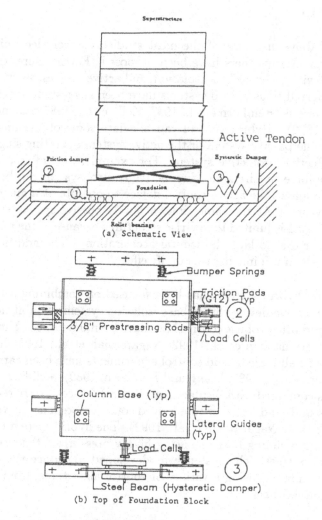

(a) Schematic View

(b) Top of Foundation Block

Fig. 14   Structure with Sliding Interface and Displacement Control Devices
(Wang and Reinhorn, 1989)

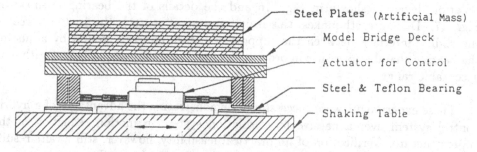

Fig. 15   Sliding Hybrid Control System (Riley et al., 1992)

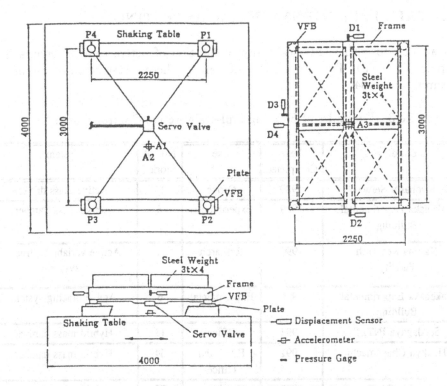

Fig. 16   Structural Model with Hybrid Sliding Isolation Device
(Feng et al., 1991)

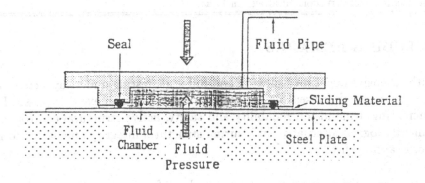

Fig. 17   Friction Controllable Sliding Bearing (Feng et al., 1991)

FULL-SCALE IMPLEMENTATION AND TESTING

As alluded to earlier, full-scale implementation of active control devices has taken place (Table 3). The performance of some of these systems will be discussed in Lectures 33 and 34.

Table 3   Examples of Existing Full-scale Active Control System*

Building	Year Completed	Use	No. of Stories	System
Kyobashi Seiwa	1989	Office	11	Active mass driver
Takenaka Experimental Building	1989	Experimental	6	Active mass damper
Kajima Research Facility	1990	Research	3	Active variable stiffness system
Takenaka Experimental Building	1990	Experimental	6	Active bracing system
Sendagaya INTES	1991	Office	11	Hybrid mass damper
Hankyu Chayamachi	1992	Hotel and Office	34	Hybrid mass damper
MM21	1993	Hotel and Office	70	Hybrid mass damper
Long Term Credit Bank	1993	Office	21	Hybrid mass damper
KS Project	1993	Multiple Use	29	Hybrid mass damper
Ando Nishikicho	1993	Office and Residence	14	Hybrid mass damper

*All steel frame construction; all located in Japan

CONCLUDING REMARKS

With extensive experimental work and full-scale testing underway, active structural control research for seismic applications has entered an exciting phase. Faced with increasing demands on reliability and safety, active structural control can be an eminently logical alternative in insuring structural integrity and safety to more traditional approaches.

At the same time, however, a large number of serious challenges remain and they must be addressed before active structural control can gain general acceptance by the civil engineering and construction professions at large. Some of these issues are discussed below.

Capital Cost and Maintenance

Maintenance is certainly necessary for active systems and this is an important issue particularly due to the fact that, when active control is only used to counter large seismic and other environmental forces, it is likely that the control system will be infrequently activated. The reliability of a system operating largely in a standby mode and the related problems of maintenance and performance qualification become an important issue.

Cost, however, is not likely to be an obstacle. Recent phenomenal advances in allied technology such as computers, electronics and instrumentation all reflect favorably on the cost factor. Based on recent experiences in the fabrication of full-scale systems, active systems can in fact be more economical when used in strengthening existing structures than, for example, the use of base isolation systems. This is largely due to the fact that active systems can be designed such that they are not structurally invasive. More studies, however, are needed to address the cost issue in more concrete terms.

Reliance on External Power and Reliability

Active systems rely on power sources and, when these sources in turn rely on all the support utility systems, this power dependence on the part of an active system presents serious challenges since the utility systems, unfortunately, are most vulnerable at the precise moment when they are most needed. The scope of the reliability problem is thus considerably enlarged if all possible ramifications are considered.

It should be noted that this problem has been addressed in the design of the full-scale active bracing system discussed in the preceding section. Since the control interval for earthquake-excited motions is of the order of one minute or less for each episode, the power requirement of this system is such that it can be supplied by currently available accumulators. This design strategy would eliminate its dependence on external power at the time of control execution.

On reliability, not to be minimized is the psychological side of this issue. There may exist a significant psychological barrier on the part of the occupants of a structure in accepting the idea of an actively controlled structure, perhaps leading to perceived reliability-related concerns.

Nontraditional Technology

Since the concept of an active-controlled structure is a significant departure from traditional structural concepts, obstacles exist with respect to its acceptance by the civil engineering and construction profession at large. This is particularly true when structural safety is to rely upon an active control system. More full-scale demonstration projects are thus needed for purposes of concept verification and education.

System Robustness

As demanded by reliability, cost and hardware development, applicable active control systems must be simple. Simple control concepts using minimum number of actuators and sensors may well deserve more attention in the near future. Simple control, of course, does not mean simple problems. Since civil engineering structures are complex systems, this inherent incompatibility gives rise to a number of challenging problems from the standpoint of system robustness, controllability and effectiveness.

Active vs. Passive Control

While some progress has been made in this direction, more comprehensive studies are certainly needed in order to realistically evaluate the relative merits of alternative structural protection techniques on the basis of practical criteria such as performance, structural type, site characteristics and cost-effectiveness. However, to find answers to these questions are more long-term tasks since they will depend on specific structural applications, hardware details and a variety of other issues, many of which need to be better understood and further developed.

REFERENCES

Agababian Assoc., (1984a). *Validation of Pulse Techniques for the Simulation of Earthquake Motions in Civil Structures*, AA Rept. No. R-7824-5489, El Segondo, CA.

Agababian Assoc., (1984b). *Induced Earthquake Motion in Civil Structures by Pulse Methods*, AA Rept. No. R-8428-5764, El Segondo, CA.

Chung, L.L., Reinhorn, A.M. and Soong, T.T., (1988). "Experiments on Active Control of Seismic Structures," *ASCE J. Eng. Mech.*, Vol. 114, pp. 241-256.

Chung, L.L., Lin, R.C., Soong, T.T. and Reinhorn, A.M., (1989). "Experimental Study of Active Control of MDOF Seismic Structures," *ASCE J. Eng. Mech.*, Vol. 115, pp. 1609-1627.

Feng. Q., Shinozuka, M., Fujii, S. and Fujita, T. (1991), "A Hybrid Isolation System for Bridges," *Proc. 1st US-Japan Workshop on Earthquake Protection Systems for Bridges*, Buffalo, NY.

Inaudi, J.A., Lopez Almansa, F., Kelly, J.M. and Rodellar, J. (1992), "Predictive Control of Base Isolated Structures," *Earthquake Engrg. Struct. Dyn.*, Vol. 21, 471-482.

Inaudi, J.A., Kelly, J.M. and Pu, J.P. (1993), "Optimal Control and Frequency Shaping Techniques for Active Isolation," *Proc of ATC 17-1 Seminar on Seismic Isolation, Passive Energy Dissipation, and Active Control*, San Francisco, Vol. 2, 787-798.

Kelly, J.M., Leitmann, G. and Soldatos, A.G., (1987). "Robust Control of Base-Isolated Structures Under Earthquake Excitations," *J. Optim. Th. Appl.*, Vol. 53, pp. 159-180.

Miller, R.K., Masri, S.F., Dehghanyar, T.J. and Caughey, T.K., (1987). "Active Vibration Control of Large Civil Engineering Structures," *ASCE J. Eng. Mech.*, Vol. 114, pp. 1542-1570.

Nagarajaiah, S., Riley, M., Reinhorn, A. and Shinozuka, M. (1992), "Hybrid Control of Sliding Isolated Bridges," *Proc. 1992 Pressure Vessels and Piping Conf.*, ASME/PVP-237, Vol. 2, 83-89.

Nagarajaiah, S., Riley, M. and Reinhorn, A. (1993), "Control of Sliding Isolated Bridge with Absolute Acceleration Feedback," *ASCE J. of Engrg. Mech.*, Vol. 119(10) (in print).

Reinhorn, A.M., Soong, T.T. and Wen, C.Y., (1987). "Base Isolated Structures with Active Control," *Proc. ASME PVD Conf.*, PVP-127, pp. 413-420, San Diego, CA.

Reinhorn, A.M., Soong, T.T., et al., (1989). *1:4 Scale Model Studies of Active Tendon Systems and Active Mass Dampers for Seismic Protection*, Tech. Rep. NCEER-89-0026, National Center for Earthquake Engineering Research, Buffalo, NY.

Riley, M.A., Nagarajaiah, S. and Reinhorn, A.M. (1992), "Hybrid Control of Absolute Motion in Aseismically Isolated Bridges," *Proc. Third NSF Workshop on Bridge Engineering Research*, Univ. of California, San Diego, 239-242.

Riley, M.A., Subramaniam, R., Nagarajaiah, S. and Reinhorn, A.M. (1993), "Hybrid Control of Sliding Base Isolated Structures," *Proc of ATC 17-1 Seminar on Seismic Isolation, Passive Energy Dissipation, and Active Control*, San Francisco, Vol. 2, 799-810.

Roorda, J., (1980). "Experiments in Feedback Control of Structures," *Structural Control*, (H.H.E. Leipholz, ed.), North Holland, Amsterdam, pp. 629-661.

Soong, T.T. and Skinner, G.T., (1981). "Experimental Study of Active Structural Control," *ASCE J. Eng. Mech., Div.*, Vol. 107, pp. 1057-1068.

Subramaniam, R., Reinhorn, A.M. and Nagarajaiah, S. (1992), "Application of Fuzzy Set Theory to the Active Control of Base Isolated Structures," *Proc. US/China/Japan Trilateral Workshop on Structural Control*, China, 153-159.

Traina, M.I., Masri, S.F., et al., (1988). "An Experimental Study of Earthquake Response of Building Models Provided with Active Damping Devices," *Proc.*

Ninth World Conference on Earthquake Engineering, Vol. VIII, pp. 447-452, Tokyo/Kyoto, Japan.

Wang, Y.P. and Reinhorn, A.M. (1989), "Motion Control of Sliding Isolated Structures," *Seismic, Shock, and Vibration Isolation - 1989* (Eds. H. Chung and T. Fujita), ASME/PVP Vol. 181, 89-94.

Yang. J.N., Li, Z., Danielians, A. and Liu, S.C. (1992a), "Aseismic Hybrid Control of Nonlinear and Hysteretic Structures I," *ASCE J. Engrg. Mech.*, Vol. 118(7), 1423-1440.

Yang, J.N., Li, Z. and Liu, S.C. (1992b), "Stable Controllers for Instantaneous Optimal Control," *ASCE J. Engrg. Mech.*, Vol. 118(8), 1612-1630.

Yang, J.N., Li, Z and Vongchavalitkul, S. (1993), "Nonlinear Control of Buildings using Hybrid Systems," *Proc of ATC 17-1 Seminar on Seismic Isolation, Passive Energy Dissipation, and Active Control*, San Francisco, Vol. 2, 811-822.

CHAPTER XVI

ACTIVE CONTROL DEVELOPMENT IN THE U.S. AND CASE STUDIES

T.T. Soong

State University of New York at Buffalo, Buffalo, NY, USA

ABSTRACT

As we have seen, full-scale implementation of active control devices in buildings has taken place, but all installations in either full-scale test structures or new buildings can only be found in Japan. However, some of the operational full-scale active systems are the result of US-Japan research collaborations and U.S. researchers were responsible for the design, fabrication and installation of an active bracing system in a full-scale dedicated test structure for structural response control under seismic loads (Soong et al., 1991; Reinhorn et al., 1992; Reinhorn et al., 1993). In addition, at least two designs have been completed for the purpose of retrofitting existing deficient structures in the U.S. The active bracing system and one of the design projects are briefly described below.

FULL-SCALE ACTIVE BRACING SYSTEM

In 1990, a full-scale active bracing system (ABS) was designed and fabricated by U.S. researchers and installed in a dedicated test structure in Tokyo for performance verification of the system under actual seismic ground motions. As shown in Fig. 1, the structure is a symmetric two-bay six-story building constructed of rigidly connected steel frames of rectangular tube columns and W-shaped beams with reinforced concrete slabs at each of the floors. Weighing 600 metric tons, the structure was designed as a relatively flexible structure with a fundamental period of 1.1 sec in the strong direction and 1.5 sec in the weak direction, in order to simulate a typical high-rise building. The structure has low damping in the dominant modes (between 0.5% and 1% of critical).

The ABS consists of solid diagonal tube braces attached to the first story of the building as shown in Fig. 1. The control system enables longitudinal expansion and contraction of the braces by means of hydraulic servocontrolled actuators attached along the braces. The control system includes also a hydraulic power supply, an analog and digital controller, and analog sensors as shown schematically in Fig. 2.

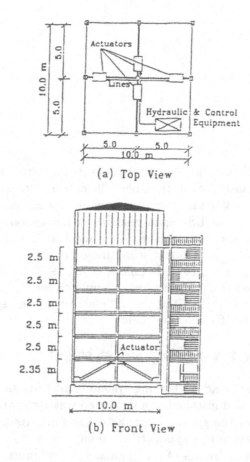

(a) Top View

(b) Front View

Fig. 1   Full-Scale Structure with ABS

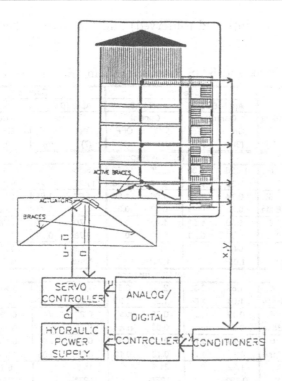

Fig. 2   Block Diagram of ABS

Since velocity sensors are available at the first, third, and sixth floors, a simplified control law was formulated using direct three-velocity feedback. Elaborate fail-safe measures were incorporated into the control software and hardware. Besides the normal operating routines, the control program contains several routines for testing and verifying integrity of the system. When a problem is detected anywhere in the system hardware or software, the control program will determine if the error is critical or not. If the error can be identified as non-critical, an attempt is made to correct the problem. If the program is unable to correct the error, or if an excessive number of correctable errors occur, the system is immediately halted through an emergency shutdown.

The structure has been subjected to several recent earthquake motions and its response was recored while the control system was automatically activated during each of these episodes. The ground motion was simultaneously recorded and was used with an analytical model to estimate the probable response of the structure in the uncontrolled mode. The peak and RMS responses at several locations of the structure and the control force requirements during two recent earthquakes are summarized in Table 1. Some typical response time histories of the sixth floor are shown in Fig. 3 for the controlled (observed) and the uncontrolled (estimated) cases.

Table 1   Peak and RMS Responses during Earthquakes

Response	Floor	X Direction				Y Direction			
		April 10,92		April 14,92		April 10,92		April 14,92	
		Control		Control		Control		Control	
		ON	OFF	ON	OFF	ON	OFF	ON	OFF
(1)	(2)	(3)	(4)	(5)	(6)	(7)	(8)	(9)	(10)
(a) Peak response									
Acceleration	6	11.36	15.30	15.25	19.50	15.66	15.46	11.50	15.03
(cm/sec/sec)	Base	7.36	7.36	8.93	8.93	11.77	11.77	9.24	9.24
Velocity	6	0.74	1.16	1.06	1.66	1.27	1.55	1.53	1.58
(cm/sec)	Base	0.23	0.23	0.37	0.37	0.74	0.74	0.49	0.49
Displacement	6	1.00	1.51	1.45	1.97	1.70	1.70	1.96	2.44
(mm)	Base	0.30	0.30	0.40	0.40	0.78	0.78	0.55	0.55
Control Force (kN)		64.0		82.7		143.2		88.6	
Control Force/Weight		1.06%		1.38%		2.38%		1.48%	
(b) RMS response									
Acceleration	6	1.89	3.39	2.28	3.51	1.97	4.67	1.91	4.14
(cm/sec/sec)	Base	1.04	1.04	1.30	1.30	1.02	1.02	1.11	1.11
Velocity	6	0.19	n/a	0.26	n/a	0.20	n/a	0.25	n/a
(cm/sec)	Base	0.04	n/a	0.06	n/a	0.05	n/a	0.06	n/a
Displacement	6	0.29	0.77	0.42	0.69	0.33	1.08	0.46	0.86
(mm)	Base	0.11	0.11	0.12	0.12	0.06	0.06	0.08	0.08
Control Force (kN)		17.2		21.7		16.3		16.9	
Control Force/Weight		0.29%		0.36%		0.27%		0.28%	

The active bracing system produces a somewhat uniform reduction of modal responses as indicated in the transfer functions in Figure 4. The controlled response has lower peaks and wider distributions around the peak, indicating damping increase.

In order to validate the analytical procedures used for predicting actual system performance, the observed response was compared with that estimated using the time step analysis and the identified properties of the system. The control forces were estimated using the uncompensated gains. The time histories of the analytical and observed responses are compared in Fig. 5 for the April 14th earthquake. The differences in the maximum peaks are less than 10%, while the RMS differences are less than 2%. Considering inherent imperfections in the structural system, the analytical predictions seem to be adequate for interpreting the structural response and for designing new systems.

It is of interest to compare the ABS performance with that of the AMD installed in the same structure. As seen from Fig. 6, the modal response of the actively controlled structure, with AMD operating at the top, is reduced substantially in the first and second modes in the y-direction and only in the first mode in the x-direction (Aizawa et al., 1990). On the other hand, the modal response under the action of ABS, as shown in Figure 4, is reduced in all modes. This behavior

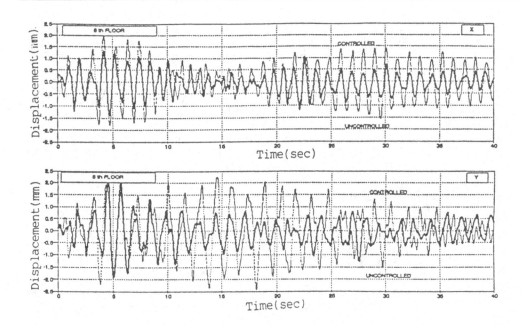

Fig. 3   Control vs Uncontrolled Response (April 14, 1992)

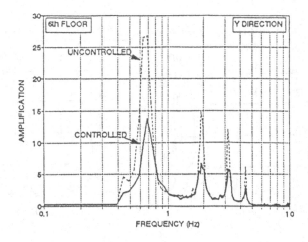

Fig. 4   Acceleration Transfer Function at Sixth Floor (April 14, 1992)

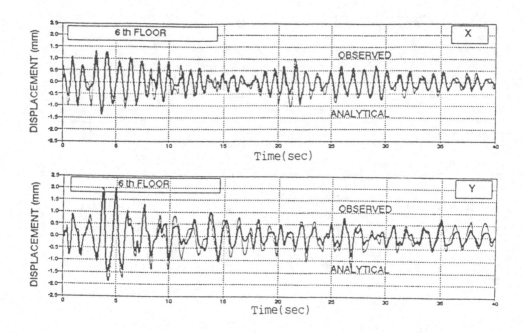

Fig. 5   Analytical vs. Observed Response (April 14, 1992)

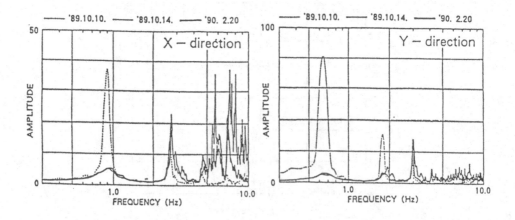

Fig. 6   Acceleration Transfer Functions With and Without Active Mass Damper
(from Aizawa et al., 1990)

was analytically predicted by Yang (1982) and observed in scaled model studies (Reinhorn, et al., 1989).

While several active mass dampers have been implemented in full-scale structures over the last few years, the active bracing system reported here represents the first full-scale active system of this type developed and tested under actual ground motions. The results obtained to date demonstrate that:

- The concept of an active tendon or bracing system, originated almost 20 years ago, has led to the successful development of the device for civil engineering structural control.

- The success of the full-scale ABS performance is the culmination of numerous analytical studies and carefully planned laboratory experiments involving model structures.

- The ABS can be implemented with existing technology under practical constraints such as power requirements and under stringent demand of reliability.

- The use of ABS in existing structures can be a practical solution for retrofit as demonstrated by this full-scale experiment. Note that the active braces were added only after the structure was completed.

- The full-scale ABS performs, by and large, as expected, and its performance can be adequately predicted through simplified analytical and simulation procedures.

- The experience gained through the development of this system can serve as an invaluable resource for the development of active structural control systems in the future.

DESIGN STUDY OF AN ACTIVE MASS DAMPER

This case study represents an interesting application of the active control concept to the strengthening of an existing building. An active mass damper was recommended in this case not because it outperforms passive devices but because of its implementability in face of practical constraints. While the design has been completed, the decision of it being implemented awaits owners' approval. The critical load in this case was wind.

The design study of a strengthening strategy for a flexible 47-story building in the U.S. began in 1990 in an attempt to determine whether the building performance under specified external loads could be significantly improved by the installation of a passive tuned mass damper system. The building considered in this study has the first natural period in excess of 10 sec, indicating an extremely flexible structure when compared to the average period in the range of 5 to 7 seconds for buildings of the same height.

The tuned mass damper (TMD) was designed and optimized. As the fourth column of Table 2 shows, significant reduction of the maximum structural response could be achieved and these reductions would meet the design objectives. However, the dynamic characteristics of the TMD itself presented great difficulties in implementation. Moreover, the required TMD stroke exceeded the space available for the system. Several iterations followed and a viable alternative for meeting both structural and control system requirements was an active mass damper (AMD) system. As column 5 of Table 2 shows, the control parameters of the AMD make it possible to be implemented. While structural response reductions are not as good as those achievable by the TMD, they fall within acceptable performance limits.

Table 2   Control Parameters of AMD and Structural Response

		W/O	TMD OPTIMAL	AMD OPTIMAL
Equiv. Struct. Properties	$\zeta(\%)$	1.0%	11.4%	6.5%
	T(sec)	10.04	10.22	10.22
Ampl. Factor	d/d_o	5.1	1.9	2.6
% Reduction of Maximum Structural Response	d	–	62.4	50.2
	v	–	64.0	53.9
	a	–	65.2	59.4
	M	–	57.9	53.7
	V	–	58.7	53.7
MD Properties	$\Delta\zeta_a + \Delta\zeta_o$	–	8%	42% + 1.3%
	T(sec)	–	10.22	9.20
Max MD Response	d(m)	–	± 5.28	± 2.54
	v(m/sec)	–	3.24	1.58
	a(g)	–	0.220	0.098

T, ζ - period and damping ratio, resp.
d,v,a - displacement, velocity, and acceleration response, resp.
M,V - overturning moment and base shear, resp.
d_o - static deflection under max pressure

The important components of the recommended biaxial AMD are similar to those required of a TMD. They include the following:

- Mass block and mass support system with pressure balanced support bearings and support bearing control system.

- Mass block restraint system, including anti-yaw torque box or equivalent hardware, anti-yaw booms, control actuator drive booms and attachments, and over-travel snubber system with reaction guides and faces.

- Biaxial motion control system fixtures, including nitrogen spring assists with a nitrogen/oil precharge control system, miscellaneous mounting structures, necessary servohydraulic control actuators with servovalves, motion and force feedback, and monitoring transducers.

- Control electronics including an oscilloscope, function generator, data acquisition and logging system, and all necessary building feedback accelerometers, error checking and interlock circuits appropriate for this type of application.

- System control center for the hydraulic power supply control, system interlocks, and a local control center with diagnostic self-test.

- Hydraulic power supply with necessary precharge pump, accumulator bank, hard lines and installation manifolds.

CONCLUDING REMARKS

As the preceding sections show, much progress has been made in research and implementation of structural control technology in the U.S. However, a number of impediments still exist. More notable is the lack of research and development expenditures by the U.S. construction industry. This is in sharp contrast to the Japanese construction industry. It is particularly evident in the development of active and hybrid control technology. Indeed, without the cooperation and financial support of the Japanese construction industry, the full-scale testing of the active bracing system would have been impossible.

It is gratifying to note that the U.S. National Science Foundation, from which much of U.S. support for structural control research has come, has recognized the importance of this research area for improvements to existing and future civil infrastructure systems. Beginning in 1989, a focused effort to foster coordinated multidisciplinary research and development in the U.S. was made. Funding was provided for creating a U.S. Panel on Structural Control Research. In 1991, the National Science Foundation established a research initiative for structural control for safety, performance and hazard mitigation. A five-year program was launched with a budget of one million dollars per year. The goal of the program was to fund research for developing control systems, robots, actuators, sensors, and energy absorbers for structures, and to investigate practical designs, fabrication, and installation techniques for field applications.

REFERENCES

Aizawa, S. Hayamizu, Y., Higashino, M., Soga, Y., Yamamoto, M. and Haniuda, N. (1990), "Experimental Study of Dual Axis Active Mass Damper," *Proceedings of the U.S. National Workshop on Structural Control Research*, (Housner, G.W. and Masri, S.F., eds), University of Southern California, Los Angeles, 68-72.

Reinhorn, A.M. and Soong, T.T., et al. (1992), *Active Bracing System: A Full-scale Implementation of Active Control*, Report NCEER-92-0020, National Center for Earthquake Engineering Research, Buffalo, NY.

Reinhorn, A.M., Soong, T.T. et al. (1993), "Full Scale Implementation of Active Control - Part II: Installation and Performance," *J. Struct. Engrg.*, ASCE, Vol. 118(6), in print.

Soong, T.T., Reinhorn, A.M., Wang, Y.P. and Lin, R.C. (1991), "Full Scale Implementation of Active Control - Part I: Design and Simulation," *Journal of Structural Engineering*, ASCE, Vol. 117(11), 3516-3536.

Yang, J.N. (1982), "Control of Tall Buildings under Earthquake Excitations," *Journal of Engineering Mechanics Division*, ASCE, Vol. 108(1), 50-68.

CHAPTER XVII

ACTIVE AND HYBRID CONTROL DEVELOPMENT IN JAPAN
- EXPERIMENTS AND IMPLEMENTATION -

H. Iemura
Kyoto University, Kyoto, Japan

ABSTRACT

In this chapter, recent developments in active and hybrid control techniques is reviewed by introducing experimental projects and some implementations being conducted in Japan. Active and hybrid control techniques are categorized into AMD (active mass damper), ATMD (active tuned Mass Damper), AB (active brace), AVS (active variable stiffness), AVD (active variable damper) and ABI (active base isolation). Shaking-table tests of flexible frame models with TMD, AMD, and ATMD at Kyoto University are also introduced.

INTRODUCTION

With vast improvement in recent construction techniques and materials, light-weight and flexible structures such as high-rise buildings and long-span bridges have been designed and constructed in Japan. It is becoming critically important to suppress dynamic response of structures due to wind and earthquake excitations, not only for their safety but also their serviceability. [Kobori et al, 1990]

To reduce dynamic response of flexible structures, passive type control techniques such as tuned mass dampers have been used. Merit of passive control method is that no external supply of energy is needed, making the structure dynamically stable and maintenance-free. Demerit is that passive control is effective only for narrow-banded frequency range and not so much effective for transient vibration due to nonstationary excitation.

Due to recent development of sensoring and digital control techniques, active and hybrid control   methods of dynamic response of structures are developed and some are implemented to buildings and bridges. Merit of active control method is that they are effective for a wide-frequency range and also for transient vibration. However, active control method needs a large amount of external energy supply and also a high level of maintenance. Hybrid control method which consists of both passive and active devices has been proposed utilizing the merits of both passive and active methods and avoiding the demerits of these methods. [Iemura et al,   1992]

In the first part of this paper, recent development in active and hybrid control techniques is reviewed by introducing experimental projects and some implementations being conducted in Japan. In the latter part, efficiency of TMD, AMD and ATMD is examined by shaking table tests of flexible structural models.

Active Mass Damper (AMD)

Control of structural response using an active-mass damper-type system provides a control force to the structure in order to reduce the response of the structure to earthquakes and strong winds by operating an auxiliary mass installed in the structure by means of actuators. There are various systems proposed, and some of them have already been implemented in actual buildings.

Kyobashi Seiwa Buildings was built in 1989 and is the first buildings in the world to have an active control system [Kobori et al, 1990]. An active mass driver system was installed to suppress dynamic response caused by earthquakes and strong winds (Fig.1). It was reported that the building had experienced several moderate earthquakes and strong winds in which ground accelerations, wind velocities and structural responses had been measured [Koshika et al 1992]. The measured responses during the earthquakes are compared with the simulated responses by numerical analyses for an uncontrolled structure. Wind response observations were performed every 30 minutes with and without control. From these comparisons, remarkable reductions in response amplitude due to the active mass driver system have been confirmed (Fig.2).

An active mass damper system which could operate in two horizontal directions had been installed in a 6-story experimental building [Aizawa et al 1990]. Two actuators set in two horizonal directions operate a 6-ton mass (Fig.3). Extensive experiments had been carried out to confirm efficiency of the control system. In addition, an active bracing system was installed at first story of this building and observation of earthquakes is being conducted.

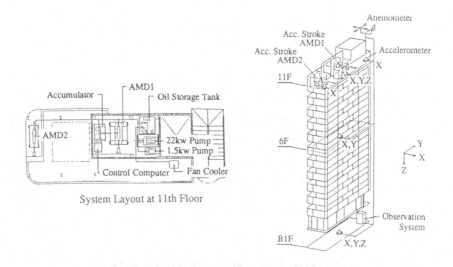

Fig.1   Active mass driver system [Kobori et al, 1990]

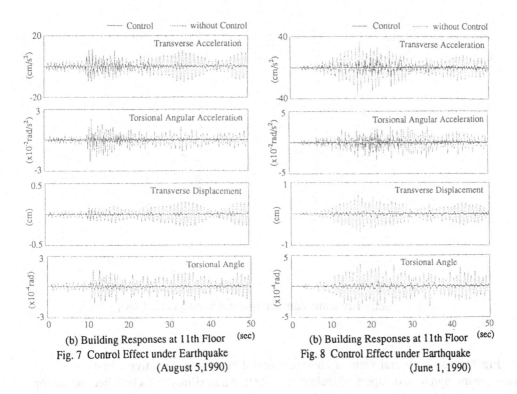

(b) Building Responses at 11th Floor   (sec)
Fig. 7 Control Effect under Earthquake
(August 5,1990)

(b) Building Responses at 11th Floor   (sec)
Fig. 8 Control Effect under Earthquake
(June 1, 1990)

Fig.2   Response with/without control using active mass driver system

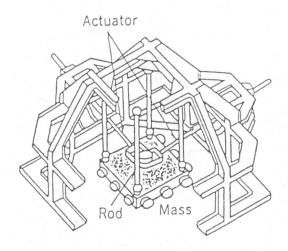

Fig.3   Active mass damper system [Aizawa et al, 1990]

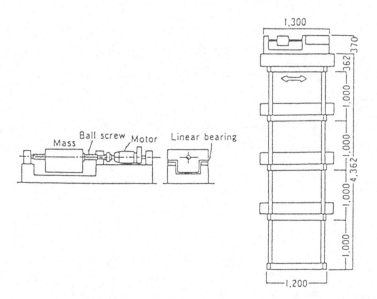

Fig.4   Active dynamic damper system [Suzuki et al, 1990]

Fig.4 shows a general view of an experimental version of an active vibration control system using dynamic dampers [Suzuki et al, 1990]. An auxiliary mass installed on the top floor of a building model is driven by an actuator. An active dynamic damper has a ball-screw driving system equipped with an AC servo-motor.

Active Tuned Mass Damper (ATMD)

In order to increase energy efficiency of actively generated control force, combined use of tuned mass damper and active control force is recently becoming popular. In addition to vibration reduction with passive-type TMD, active control force helps to reduce transient and higher-mode vibration of structures. Because of combined use of passive and active control forces, this ATMD is also called a Hybrid Control System or Hybrid Damper.

A pendulum-type tuned active mass damper has been developed (Fig.5) [Matsumoto et al, 1990]. The system has a multi-stage suspended damper mass which can be housed in a single story of a high-rise building. The hanger of the mass is winded to increase natural period of the pendulum and also to save space as shown in the figure. When sensors detect sway vibration, the computer controls servo-motors driving ball-screws to position the damper mass. It has been installed on the top floor of a 70-story building (296 meter high) and also on the top of the tower of the Akashi Kaikyo Bridge (300 meter high).

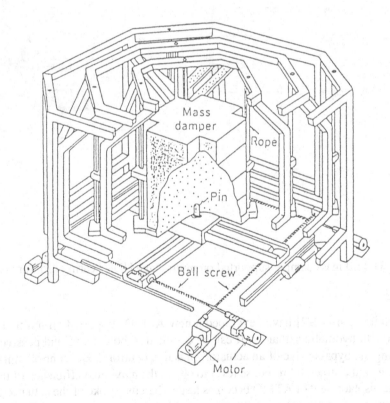

Fig.5   Pendulum-type tuned active mass damper [Matsumoto et al, 1990]

A hybrid mass damper using tuned mass damper supported on multi-stage high damping rubber bearings and an actuator has been developed (Fig.6) [Tamura et al 1992]. The significant merit of using high-damping rubber bearings is that they can take up the roles of spring, viscous damper, and also supporting guide for movement. The actuator is composed of AC servo-motors and screws. The system has been mounted on a tower-type 7 story steel frame building and verification tests are being carried out. This system is applied to a 40-story building in Osaka to reduce wind vibration.

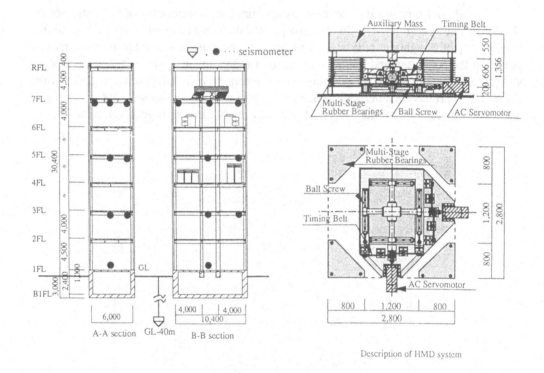

Fig.6   Hybrid mass damper supported on multi-stage HDR bearing [Tamura et al, 1992]

Fujita et al [1992] have developed a new ATMD supported on multi-stage rubber bearings with hydraulic actuators. It can be operated in both active and passive modes by controlling the bypass valve of an actuator. In active control mode, control gain is adjusted depending on the structural response level to obtain the maximum efficiency of the actuator. When the displacement of ATMD becomes larger than the stroke of the actuator due to very strong wind or earthquakes, the system will be operated in pure TMD mode. In implementation of this system to actual high-rise building, heat storage tanks are used as the mass of TMD as shown in Fig.7.

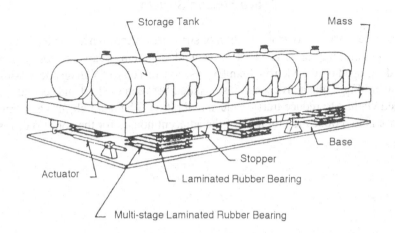

Fig.7   ATMD supported on multi-stage rubber bearing [Fujita et al, 1992]

A two-axis hybrid mass damper has been developed to reduce the vibration of tall bridge towers and high-rise building structures (Fig.8) [Tanida et al,1990]. A sliding mass shaped in an arc segment is combined with active control by an AC servo-motor. The movement of the arc segment on roller bearings is similar to that of a pendulum taking the role of a TMD. After some experiments were made with the tower structural model, it has been applied to actual bridge towers and in high-rise buildings.

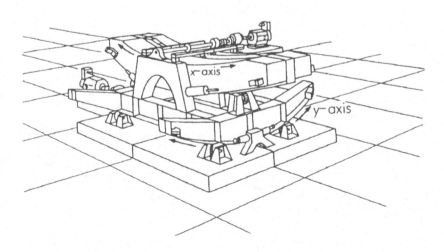

Fig.8   A two-axis hybrid-type mass damper [Tanida et al, 1990]

Active Tendon System

The most direct way to control vibration of structures is through active tendon or active brace installed in the structures. The first active tendon system in Japan to control structural vibration is developed by the Metropolitan Expressway in Tokyo [Yahagi and Yoshida, 19 85]. As shown in Fig.9, the active tendon is installed between the first and second stories of a steel framed viaduct to reduce traffic induced vibration. The main objective of this active tendon control is to reduce the vibration of the adjacent houses for the comfort of residents nearby.

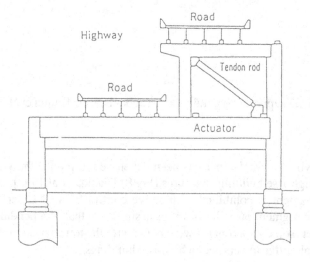

Fig.9   Active tendon system for highway viaduct [Yahagi and Yoshida, 1985]

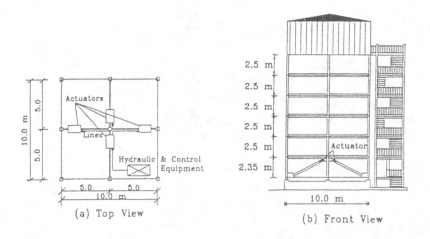

Fig.10   Active brace control system for buildings [Reinhorn and Soong, 1990]

Takenaka Construction company in Japan and the State University of New York at Buffalo had a joint research project on active brace system for buildings. The system was experimentally tested on a full-scale structure as shown in Fig.10 [Reinhorn and Soong,199 0]. A biaxial bracing system was assembled in the middle frames in two orthogonal directions. The bracing system consisted of two diagonal circular tubular braces attached within the first floor in each bay of the frame. A hydraulic actuator 312 kN capacity is assembled in each brace at the connection to the center column of the structure.

Active Variable Stiffness (AVS) and Active Variable Damper (AVD) Control Systems

One of the significant demerits of active control of structures is the high amount of energy required to reduce vibration of large structural systems. AVS and AVD control systems are developed to overcome this demerit. In AVS and AVD, stiffness and damping capacity are changed by merely adjusting the valves in oil circuits. Hence, almost no energy is required; but, on the other hand, sophistication in control algorithm becomes necessary.

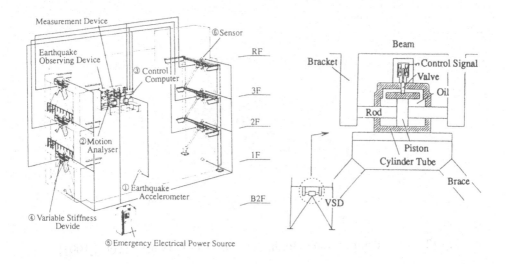

Fig. 11 Active variable stiffness system for buildings [Kobori et al, 1990]

In nonresonant-type controlled structure, the system actively controls the vibration characteristics of a structure so that resonance with input motion can be avoided and the response can be suppressed. To achieve this objective, the active variable stiffness system has been developed and applied to a 3-story steel building [Kobori et al, 1990]. Braces have been placed in the transverse direction of the building, and the variable stiffness device has been installed between the top of brace and the upper beam. The earthquake motion analysis type of feedforward control scheme has been adopted. Depending on the predominant frequency of the excitation in previous cycle of structural response, 3 types of structural stiffness (rigid, medium, soft) are selected to reduce the dynamically amplified response.

A variable damper is being developed at the Public Works Research Institute (Fig.12) [Kawashima et al, 1991]. The viscous damping force of this damper is varied depending on the response of highway bridges. Damping coefficient of the damper takes large values for a small amplitude as a damper stopper for traffic and wind loads. As amplitude becomes larger due to earthquake excitations, the damping coefficient is decreased so that energy dissipation is made optimum and inertial force adjusted appropriately. For excessive amplitude, the damping coefficient is increased again to suppress the response.

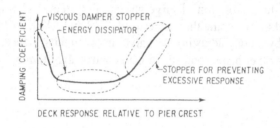

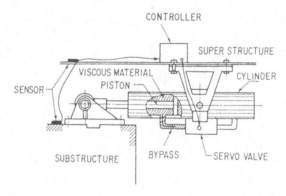

Fig.12 Active damper system for highway bridges [Kawashima et al,1991]

Active Base Isolation

"Super earthquake free control system" for building is investigated by a research group at the Technical Research Institute of Obayashi Corporation [Kageyama et al, 1992]. This system consists of passive-type base isolation bearings of laminated rubber and active-type hydraulic actuators attached between the foundation and the ground as shown in Fig.13. The target of this research is to develop the system which gives the perfect isolation of the structure from the ground.

Shinozuka, Fujii, Feng et al [1992] have developed the friction-controlled sliding bearings. The friction force is controlled by adjusting the normal axial force by the oil pressure as shown in Fig.14. By controlling the friction force, the sliding displacement can be confined within an acceptable range, while keeping the overall isolation performance optimal under the circumstances. A model hybrid system has been tested on a shaking table.

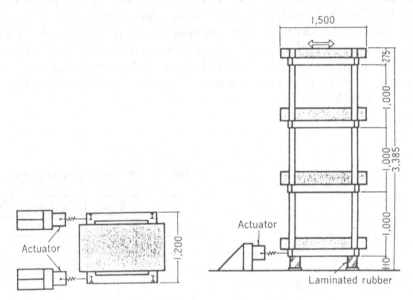

Fig.13 Super earthquake-free control system [Kageyama et al, 1992]

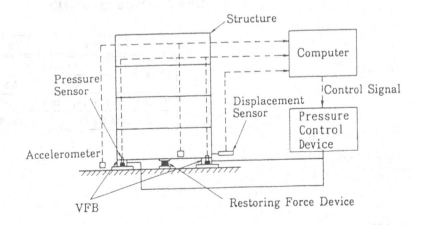

Fig.14 Friction-controllable sliding bearing system [Shinozuka et al, 1992]

Shaking Table Tests of a Flexible Structural Model with TMD, AMD and ATMD at Kyoto University

For practical implementation of TMD, AMD and ATMD for structures, it is important to find efficiency of each control system for random excitations.The author made analytical and numerical studies on the efficiency of different control methods.[Iemura et al, 1992] To verify the results , a 3-DOF structural frame model shown in Fig.15 with and without control devices is tested on a shaking table at Kyoto University. Natural frequency and participation factors of each mode of the model is shown in Table 1. Mass, stiffness and damping ratio of the TMD shown in Table 2 are used for the experiment. The mass of the TMD consists of the AC servo-motor, moving mass, driving guides, and velocity meter. At the time of the experiment of the TMD, moving mass is fixed and the TMD is hanged from the third floor. In order to work as the hybrid type ATMD, moving mass is driven by the motor. For the pure active control experiment, the motor and the moving mass are set directly on the third floor.

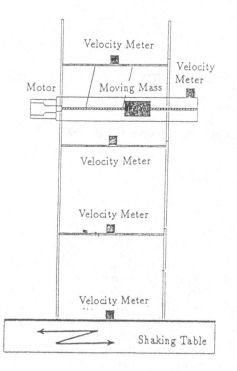

Table 1   Natural frequency of the model

Mode	T (sec)	Participation Factor
1	0.6578	1.2204
2	0.2580	0.3493
3	0.1568	-0.1341

Table 2   Properties of TMD

Parameter		Value
Mass (kg)	Moving Mass	3.5
	Others	5.5
Spring Constant (kgf/cm)		0.581
Damping Ratio (%)		25.06

Fig.15 3-DOF model with TMD/ATMD

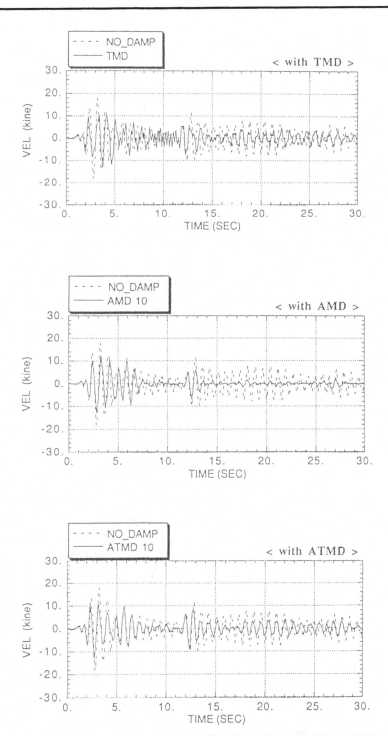

Fig.16   Velocity response of 3rd story of the model subjected to El Centro NS Component

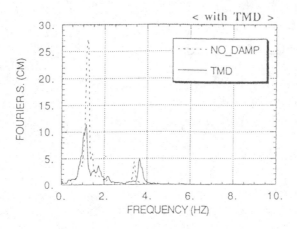

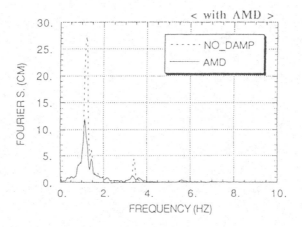

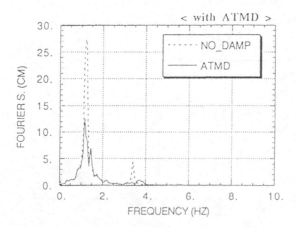

Fig.17 Fourier spectrum of 3rd story response

Fig.16 shows dynamic response of the model with and without control devices, subjected to the N-S component of El Centro acceleration records obtained during the Imperial Valley Earthquake in 1940. Velocity response of third floor of the models is compared with and without dampers. Dotted line shows the response without any damper. Active control force is adjusted to give similar level of response of the model with TMD.

Fig.17 shows fourier spectrum of the response shown in Fig.16. It is clearly found that the second mode response is not reduced by the TMD but effectively reduced by AMD and ATMD. TMD is effective only in the first mode frequency range, while active control force can cover wide frequency range.

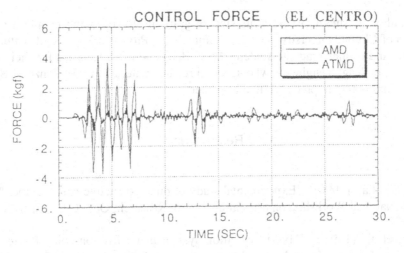

Fig.18(a) Time history of control force

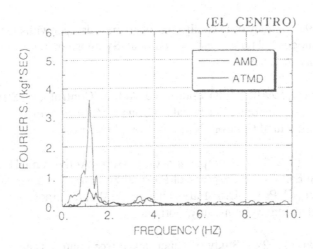

Fig.18(b) Fourier spectrum of control force

Fig.18(a) and (b) show time history and fourier spectrum of the control force of AMD and ATMD. It is found that control force of ATMD is much lower that of AMD, especially in the first mode frequency range. There is not much difference in the second mode frequency range. From the results shown in Figs.16, 17, 18, it is found that the function of the ATMD comes both from TMD and AMD. The first mode response is reduced mainly by TMD and the second mode response is reduced by AMD. This result verifies the energy efficiency of ATMD. This is the reason that the concept of ATMD is now popularly used for practical application.

Acknowledgements

The author wishes to express his sincere gratitude to following people. Professor Y.Yamada of Chubu University for his general advice, Professor K.Izuno of Ritumeikan University and former graduate students Mr.K.Baba and Mr.S.Kenzaka for their help in experiments, colleague engineers who supplied related materials, Dr.W.Tanzo of Saitama University for his help in preparation of his paper.

References

Aizawa, S., et al (1990). "Experimental study of dual axis active mass damper," Proc., U.S. National Workshop on Structural Control Research, Oct., U.S.C., Los Angeles.

Fujii, S., et al (1992). "Hybrid isolation system using friction-controllable sliding bearings," Trans., Japan National Symposium on Active Structural Control, March. (in Japanese)

Fujita, T., et al (1992). "Study of biaxially controlled mass damper with convertible active and passive modes," Trans., Japan National Symposium on Active Structural Control, March. (in Japanese)

Iemura, H., et al (1990). "Phase-adjusted Active Control of Structures with Identification of Random Earthquake Ground Motion ," Proc. U.S. National Workshop on Structural Control Research, Nov., Los Angeles.

Iemura, H., et al (1990). " Comparison of Passive, Active and Hybrid Control Techniques on Earthquake Response of Flexural Structures - Numerical Simulations and Experiments," Proc., U.S.-Italy-Japan Workshop/Symposium on Structural Control and Intelligent Systems, July, Sorrento.

Kageyama, M., et al (1992). "Study on super quake free control system of buildings," Trans. Japan National Symposium on Active Structural Control, March. (in Japanese)

Kawashima, K., et al (1991). "Current research efforts in Japan for passive and active control of highway bridges against earthquakes," Proc. 23rd Joint Meeting, U.S.-Japan Panel on Wind and Seismic Effects, UJNR, May, Tsukuba.

Kitamura, H., et al (1992). "Design and analysis of a tall building with active mass damper," Proc. 10WCEE, Madrid, Spain.

Kobori, T., et al (1990). "State-of-the-Art of Seismic Response Control Research in Japan," Proc. U.S. National Workshop on Structural Control Research, Nov., Los Angeles.

Kobori, T., et al (1990). "Study on active mass driver (AMD) system - active seismic response controlled structure," 4th World Congress of Council on Tall Buildings and Urban Habitat, Nov., Hongkong.

Kobori, T., et al (1990). "Experimental study on active variable stiffness system - active seismic response controlled structure," 4th World Congress of Council on Tall Buildings and Urban Habitat, Nov., Hongkong.

Koshika, N., et al (1992). "Control effect of active mass driver system during earthquakes and strong winds," Trans. Japan National Symposium on Active Structural Control, March (in Japanese)

Matsumoto, T., et al (1990). "Study on powered passive mass damper for high-rise building," Proc. AIJ Annual Meeting, Oct. (in Japanese)

Reinhorn, A. and Soong, T.T. (1990). "Full scale implementation of active bracing for seismic control of structures," Proc. U.S. National Workshop on Structural Control Research, Oct., U.S. C., Los Angeles.

Suzuki, T., et al (1990). "Active vibration control for high-rise buildings using dynamic vibration absorber driven by servo motor," Proc. U.S. National Workshop on Structural Control Research, Oct., U.S. C., Los Angeles.

Tamura, K., et al (1992). "Study on application of hybrid mass damper system to a tall building," Trans., Japan National Symposium on Active Structural Control, March. (in Japanese)

Tanida, K., et al (1990). "Development of hybrid-type mass damper combining active-type with passive-type," Proc., Dynamics and Design Conference, July, Kawasaki. (in Japanese)

Yahagi, K. and Yoshida, K. (1985). "An active control of traffic vibration on the urban viaducts," Proc., JSCE, No.356/I-3, April. (in Japanese)

CHAPTER XVIII

SUMMARY AND CONCLUDING REMARKS

T.T. Soong
State University of New York at Buffalo, Buffalo, NY, USA

ABSTRACT

As we have seen, remarkable progress has been made in the area of base isolation, passive energy dissipation, and active control. Recent advances in seismic isolation hardware and the benefit offered by this technology have been responsible for a rapid increase in the number of base-isolated buildings in the world, both new construction and retrofit.

The basic role of passive energy dissipation devices when incorporated into a structure is to consume a portion of the input energy, thereby reducing energy dissipation demand on primary structural members and minimizing possible structural damage. Over the last twenty years, serious efforts have been undertaken to develop the concept of supplemental damping into a workable technology and it has now reached the stage where a number of these devices have been installed in structures throughout the world.

In comparison with base isolation and passive energy dissipation, active control is a relatively new area of research and technology development. However, we again see a rapid development in this area, to the point that active systems have been installed in several structures in Japan.

While considerable progress has been made, it is also clear that a large number of challenges remain. Base isolation, passive energy dissipation, and active control have recently been reviewed by, respectively, Kelly (1993), Hanson et al. (1993), and Soong and Reinhorn (1993). It appears appropriate to conclude this lecture series by quoting remarks made by these authors regarding areas of research needs and possible future directions.

BASE ISOLATION (KELLY, 1993)

It seems clear that the increasing acceptance of base isolation throughout the world will lead to many more applications of this technology. It is also clear that while elastomeric systems will continue to be used, there is a willingness to try other systems. The initial skepticism that was so prevalent when elastomeric systems were initially proposed is no longer evident, and the newer approaches which are currently being developed will benefit from this more receptive climate and lead to the development of systems based on different mechanisms and materials.

For all systems, the most important area for future research is that of the long-term stability of the mechanical characteristics of the isolator and its constituent materials. The long-term performance of isolators can best be developed from inspection and retesting of samples that have been in service for many years. Elastomeric systems in the form of non-seismic bridge bearings have been used for upwards of thirty years and a record of satisfactory performance has been established (Stevenson 1985; Taylor et al. 1992).

Many of the completed base-isolated buildings have experienced earthquakes and so far their performance has been as predicted. The earthquakes, if close, have been small or have been moderate and distant so that the accelerations experienced have not been large. As more isolated buildings are build in earthquake-prone regions of the world, we can anticipate learning more about the behavior of such structures and it will be possible to reduce the degree of conservatism that is currently present in the design of these structures. It should be possible to bring about an alignment of the codes for fixed-base and isolated structures and have a common code based on the specified level of seismic hazard and structural performance and in this way allow the economic use of this new technology for those building types for which it is appropriate.

It is clear that the use of seismic isolation has finally achieved a level of acceptance that will ensure its continued use and its further development and that this new and radical approach to seismic design will be able to provide safer buildings at little additional cost as compared to conventional design. Additionally, base isolation may play a major role in the future in projects as diverse as advanced nuclear reactors and public housing in developing countries.

PASSIVE ENERGY DISSIPATION (HANSON, ET AL., 1993)

Significant advances in the field of passive energy dissipation for improved structural seismic resistance have been made in recent years. Developments in the research and analysis arena have been paralleled by significant improvements and refinement of available device hardware. Most, if not all, of these systems are now sufficiently well understood for their use in new or retrofit design of buildings.

A wide range of behavioral characteristics are possible. Of particular importance for all energy dissipation devices is that they have repeatable and stable force-deformation behavior under repeated cyclic loading, and reliable behavior in the long term. Seismic damping devices typically possess nonlinear force-displacement behavior, and thus nonlinear time-history analysis methods are usually necessary to verify design performance, at least until the confidence level of designers increases.

As the number of viable energy dissipation systems increases, it is becoming increasingly necessary to find a common basis for evaluation and comparison of these systems and their use in established reasonable design standards. There have been few test programs which have included more than two or three systems, and those have not attempted comparisons beyond evaluation of general performance characteristics. The following conclusions are presented on the basis of the foregoing information.

1. The effects of supplemental viscous damping on the earthquake response of buildings can be considered separately from other effects, including structural ductilities, and are complementary to the current code procedures when using R or R_w factors.

2. The design coefficient for added damping building systems can be selected using the reduction factor. This requires that both the fraction of critical damping assumed for the design spectra and the equivalent viscous damping provided by the energy dissipation system be known.

3. The relative effectiveness of the damping decreases as the amount provided increases. The cost effective limit for an energy dissipation approach will depend upon the structural system and the type of damping device selected.

4. There is a need to consolidate the basis for subsequent developments and applications of seismic energy dissipation systems. Some of the current general and specific issues related to these future applications are:

   a. Stable and repeatable performance characteristics under dynamic loading.
   b. Variable performance characteristics as a function of loading, e.g., temperature, amplitude and frequency dependence.
   c. Practical design methodologies and criteria.

  d. Long term reliability, e.g., deterioration, corrosion, design life.
  e. Maintenance requirements and in situ performance evaluation.
  f. Standards for device assessment and comparison.

We are on the verge of significant growth and development in this field. The potential for improved seismic safety and cost effectiveness is enormous.

ACTIVE CONTROL (SOONG AND REINHORN, 1993)

Research needs in active and hybrid control can be grouped into the following three broad areas: basic research, system integration, and standardization.

A more pertinent issue with direct impact to structural applications is system integration. There is a lack of system integration effort which is needed in order to allow the development of rational design and performance evaluation guidelines. Research needs in this area are outlined below.

While much progress has been made in the study of components of passive and active damping systems, much more attention needs to be paid to the overall performance of integrated systems when applied to realistic structures. A structural control system, active and hybrid systems in particular, consists of a number of important components. In addition, some systems operate only intermittently with long dormant periods. Thus, a number of implementation-related issues must be addressed before these systems can be widely accepted.

A more systematic representation of an integrated system is shown in Fig. 1, representing an active bracing system. Components such as sensors, controllers, hydraulic systems, and force generators are integrated into the active system and their integrated performance produces a feedback or a feedforward operation in a direct active or a hybrid mode. Numerous publications and research efforts have been concerned with component studies; however, little attention has been paid to issues related to system integration. Table 1 indicates some of these issues related to system integration based on the schematic diagram in Fig. 1 that have not been properly addressed.

Some of the important issues needing immediate attention include:

Control Algorithm and System Optimization

Control algorithms for active systems need to be improved to include, besides the usual structural performance parameters, considerations such as parametric uncertainties, realistic control force constraints, control system/structure interactions, actuator/sensor dynamics, power source as well as other practical constraints such as spillover and time delay. Indeed, control algorithms are only one component of the system optimization issue which also includes actuator/sensor placements,

Table 1   System Integration Systems

1 - Overall System Integration	2 - Partial Integration: Structure & Sensors

Components involved:
   - (A) Excitation
   - (B) Plant
   - (C) Response
   - (D) Sensors
   - (E) System controller
   - (F) Force generator
 	
Issues:
   - System optimization
   - Integrated fail-safe operations
   - Cost-benefit analysis
   - System uncertainties
   - Integrated reliability & safety
   - Long term maintenance
   - Operational issues
   - Time delay effects
	Components involved:
	   - (B) Plant
	   - (C) Response
	   - (D) Sensors
	
	Issues:
	   - Efficiency of integration
	   - Integrated construction
	   - Self-diagnostics
	   - Remote vs. integral devices

3 - Partial Integration: System Controller	4 - Overall System Development

Components involved:
   - (D) Sensors
   - (E) Controller
   - (F) Force generator
 	
Issues:
   - Sensor/controller/actuator integration
   - Integrated control algorithm
   - Modular design
   - Operational reliability
   - Integrated power sources
	Phases:
	   - Formulation
	   - Experimentation
	   - Design issues
	   - Implementation
	
	Issues:
	   - Individual vs. overall testing
	   - Generalized guidelines & codes
	   - Prediction software
	   - Simulation guidelines
	   - Software/hardware integration

response bounds, and power requirement. This issue also requires experimental verification under realistic conditions.

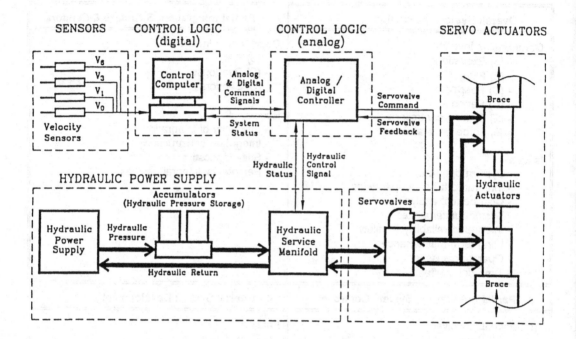

Fig. 1   Details of Active Bracing System

Integrated Fail-safe Operation

The control/structure system is expected to encounter conditions that are detrimental to their safe operation. Thus, system monitoring is necessary and, when unsafe conditions exist, the control operation can be either modified or shut down in a controlled manner in order to prevent other adverse effects. Realistic systems have components that may respond to emergency actions differently. Therefore, it is desirable to develop fail-safe operation procedures that will not endanger the control system but still provide required safety margins for the structure.

Integrated Safety, Reliability and Maintenance

The integrated control/structure system gives rise to several new concerns such as (i) improving the structural system sufficiently in terms of increasing its safety margin against damage or collapse, (ii) reliability of operation of the control system and its influence on structural components and response objectives, and (iii) maintaining reliability of controllers and controller operations by means of scheduled maintenance, self diagnostics, and automated correction. These issues can be addressed through studying realistic integrated systems comprising of all components for which safety and reliability issues are well defined. Such development must include case studies and long-term experimental verification.

Operational Issues - Software/Hardware Integration

Although the control algorithm is the centerpiece of control operation, its implementation is directly related to operational problems. The interface between operational and nonoperational stages, data transfer and storage during operation, quality of transferred data, and algorithm robustness are some of the operational issues. Integration of maintenance procedures into control operations along with self identification and diagnostics is among the most important issues for long-term operation. The solution for ensuring reliable control operation requires software/hardware integration based on a clear understanding of their interactions and ramifications.

Self Identification and Diagnostics

Sensors are not only an integral part of an active control system but also essential in structural identification and diagnostics. Integration of sensors and the structure thus becomes an important issue. In both passive and active control, this issue needs to be addressed to insure that long-term operation is not adversely affected by changes in structural behavior, either involuntary or man-made. Integration of sensors along with transmission components and logical interpretation modules (chips or computers) necessitates development of integrated procedures. These procedures can then provide self identification and adaptability of controllers as well as proper diagnostics of structural performance.

Integration of Sensors/Controllers/Force Generators with Power Source

Structural control systems packaged appropriately in the form of commonly used structural elements or simple attachments are desirable since they do not require complex on-site preparations and assemblies. As such, development of "smart" structural components seems to be an obvious solution. An active tuned mass damper in which hydraulic power supply, controllers and sensors are embedded can be such an example. A telescopic bar which houses hydraulic actuators with accumulators, logical controller and necessary sensors can be another. This type of "self-contained" systems requires a certain degree of design flexibility that allows adaptation to various structural systems under different conditions. This integration effort will involve engineering development and experimentation along with development of analytical tools that enable physical integration of devices with structural systems.

Cost Benefit Analysis - Case Studies

It is clear that one of the guiding factors in choosing a particular control strategy and hardware is the cost-benefit issue. In traditional construction, the investment cost is usually a sufficient indicator of life cycle expenses required for a structure. This is due to the fact that maintenance and replacement costs either do not differ very much from one alternative to another or are proportional to the initial investment. Moreover, serviceability and safety margins are not included in a cost-benefit analysis since they are already embedded in the initial design criteria.

However, for control/structure systems, the cost analysis needs to itemize both investment costs and maintenance requirements along with safety and service cost-benefits. Such costs need to be evaluated for both "controlled" and "uncontrolled" structures if a prudent decision is to be made. Prototype cost analyses need to be developed from case studies.

REFERENCES

Kelly, J.M. (1993), "State-of-the-Art and State-of-the-Practice in Base Isolation," *Proc. of ATC 17-1 Seminar on Seismic Isolation, Passive Energy Dissipation, and Active Control*, San Francisco, Vol. 1, 9-28.

Hanson, R.D., Aiken, I., Nims, D.K., Ritcher, P.I. and Bachman, R. (1993), "State-of-the-Art and State-of-the-Practice in Seismic Energy Dissipation," *Proc. of ATC 17-1 Seminar on Seismic Isolation, Passive Energy Dissipation, and Active Control*, San Francisco, Vol. 2, 449-471.

Soong, T.T. and Reinhorn, A.M. (1993), "Case Studies of Active Control and Implementational Issues," *Proc. of ATC 17-1 Seminar on Seismic Isolation, Passive Energy Dissipation, and Active Control*, San Francisco, Vol. 2, 701-714.

Stevenson, A. (1985), "Longevity of Natural Rubber in Structural Bearings," *Plastics and Rubber Processing and Applications*, Vol. 5, 253-262.

Taylor, A.W., Lin, A.N. and Martin, J.W. (1992), "Performance of Elastomers in Isolation Bearings: A Literature Review," *Earthquake Spectra*, Vol 8(2), 279-304.